Debbie Bates & Liz Kettle

Landauer Publishing

Threads
The basics & beyond

Debbie Bates & Liz Kettle

Threads The basics & beyond

This book was designed, produced, and published by Landauer Books
A division of Landauer Corporation
3100 101st Street, Urbandale, IA 50322
www.landauercorp.com   800/557-2144

President/Publisher: Jeramy Lanigan Landauer
Vice President of Sales and Operations: Kitty Jacobson
Managing Editor: Jeri Simon
Art Director: Laurel Albright
Photographer: Sue Voegtlin

ISBN 13: 978-0-9825586-1-4
ISBN 10: 0-9825586-1-9
Library of Congress
Control Number: 2009938123

This book printed on
acid-free paper.
Printed in China

10-9-8-7-6-5-4-3-2-1

Introduction

Our stitch journey began, like many journeys do, with no clear idea of where we were going. First we fell madly and passionately in love with the luscious colors and textures of exotic rayon, silk, and metallic threads. Something about threads' incredible potential to make anything drew us in. Then love waned as we became frustrated with its unruly ways and sometimes difficult disposition. However, the passion continued to smolder under the surface, fueled by the amount of money invested in these threads. We rededicated ourselves to revealing their darkest mysteries and secrets and making peace with these threads that had taken over our studios. The result of our meandering journey is a deeper understanding of thread than we ever imagined.

Our personal stitch journey has taught us that understanding our tools allows us to expand our creativity and explore the art of our craft more deeply. We are excited that you will learn about the world of thread in this book, plus gain a better understanding of your sewing machine as well as other tools of our trade. With this knowledge we know you can stitch whatever you can imagine.

We invite you to join us on this journey of thread, stitch, and discovery. We ask that you be brave as you explore this exciting world of thread. On your journey dare to experiment and even fail, give yourself permission to play and most of all have fun!

Bon Voyage,
Liz and Debbie

table of contents

Nuts and Bolts

Thread in Quilting

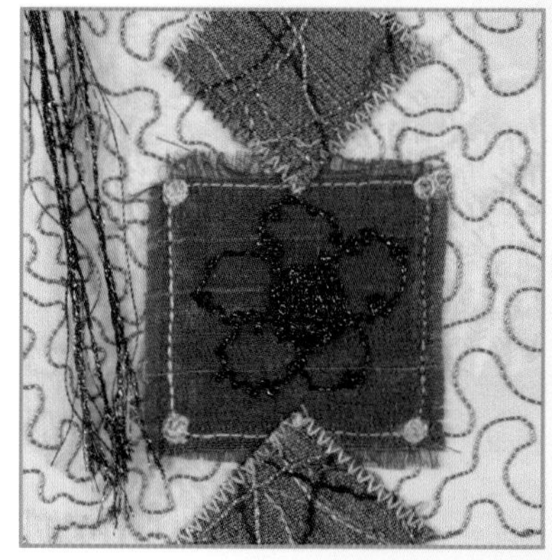

Thread as Paint

Thread as Texture

Thread as Structure

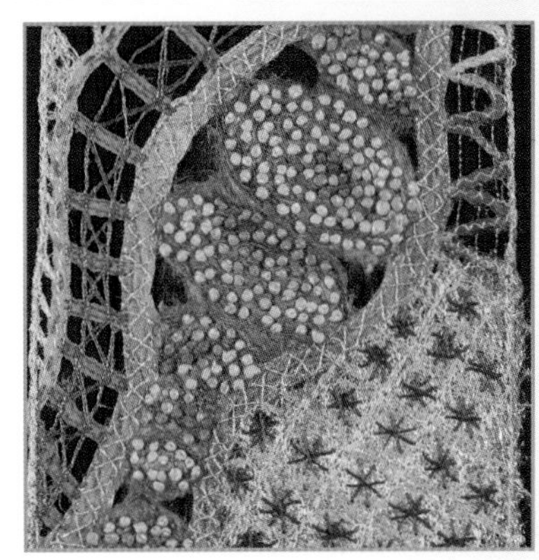

Thread as Ornament

Thread as Embellishment

the threads

Thread! *We are passionate about thread. Thread has become a specialized instrument for creativity and stitch expression. It can play a supporting, technical role or be the star of the show. Getting to know the properties of various threads will open up new possibilities for your creative work. Other than color the two primary considerations we need to think about are fiber content and thread weight.*

THREAD is defined as a fine fiber, filament, or combination of filaments twisted together and spun out to considerable length. It is generally used for binding, joining or embellishing materials versus weaving.

Thread Types, Properties and Content

POLYESTER: Polyester has a few advantages over natural fibers. It can be made into very fine strands that open a world of possibilities in micro-quilting, micro-embroidery, and working on delicate fabrics. Additionally, polyester thread is resistant to abrasion, lint free, colorfast, strong, and can be produced with a matte or shiny finish. Polyester resists shrinking, fading, and bleeding. Its elasticity and stretch recovery make it ideal for stitching stretch fabrics, knits, preshrunk fabrics, and sheers. Ultra fine and fine polyesters make excellent piecing threads, while medium weight polyesters are well suited for quilting, embroidery, construction, and embellishment.

COTTON: Mercerizing and gassing are two techniques that strengthen and de-lint cotton threads, the most popular of sewing threads. Mercerizing is a chemical treatment that prepares the cotton fibers to readily accept dye and adds strength and luster. Gassing is a process where the thread is passed through a flame to burn off excess lint and add sheen. Glazing is a hard coating applied to cotton thread to help protect it from abrasion. Glazing is primarily used in hand quilting threads. Staple refers to the length of the fiber before it is spun into thread. Long or extra long is preferred as it results in a stronger thread with less lint. Cotton thread is subject to the same shrinkage as cotton fabric and has little to no stretch, making it a favorite among quilters. It can also be considered for machine embroidery, thread painting, and lace to achieve a softer, matte look. Cotton wrapped polyester is a hybrid thread that adds the strength and elasticity of a polyester core to the softness of a cotton wrapped finish.

RAYON: Rayon is made from regenerated cellulose or wood pulp and is considered the first man-made fiber. It can hold rich dye colors, has a soft feel, and natural luster. The drawbacks of rayon are its lack of abrasion resistance which results in pilling and its lack of strength when wet.

SILK: Silk thread is made from the cocoon of the silk worm moth. Silk is prized for its strength, luminous colors, and sheen unequaled by any other thread. It is the preferred thread of couture sewing. Fine silk thread is ideal for appliqué as it practically melts into the fabric. Its strength and slick finish make silk an excellent choice for gathering or ruffling lace and fine fabrics. There are a variety of heavy weight silk embroidery threads and silk ribbons, all wonderful for hand embellishment or bobbin stitching.

WOOL: Wool thread is generally a blend of 50 percent wool and 50 percent acrylic fibers. It is a heavier thread that offers bulk, texture, and a soft finish that is excellent for appliqué, embroidery, and embellishing. Though warm and fuzzy in texture, the color range includes subtle to bold colors for con-temporary and folk-art stitch fusions. Wool threads have low tensile strength and are better suited to embellishment rather than construction. When used on a machine reduce tension to avoid breakage.

METALLIC: Metallic thread is made from an inner core of rayon, nylon, or polyester that is sometimes coated with a rice paper paste for strength and aid in binding to the outer wrapping of metal foil. Color is applied by adding a polyester film to the foil. Metallic thread weights range from fine to ultra heavy. The core of the thread affects its softness, while the color of the core material can greatly vary the look of the thread. A few tips will have you stitching metallic threads with ease:

- Use a metallic or embroidery needle.
- Sew slower.
- Reduce the top tension.
- Bypass the last thread guide, just above the needle.
- Use a thin, soft polyester thread in the bobbin.
- Some particularly wiry threads need room to relax before they encounter the tension guides. Place them on a separate spool holder behind your machine.

MYLAR/HOLOGRAPHIC THREADS: This thread is actually very thin, flat polyester film, sliced into extremely fine strips. They are quite reflective and can have a foil, pearlescent, or holographic finish. Holographic threads have a slight stretch. Treat these as metallic threads when sewing and use a fine to medium weight bobbin thread.

Specialized threads:

Our thread list would not be complete without the mention of a few specialized threads. They may not conform to the same measurement standards, but offer important contributions to your thread toolbox.

SPECIAL EFFECTS: There is a wonderful variety of thread to delight every stitcher with specialized effects. Flourescent threads add a zap of strong color, while light sensitive threads change color with light exposure. Glow in the dark threads add a surprise effect when the lights go out. Some artists are even experimenting with conductive threads. Don't be afraid to experiment and play.

MONOFILAMENT: These threads are a singular filament of polyester. Its translucent nature allows it to assume the background fabric color. It has good abrasion resistance, but is stiffer than other threads, due in part to the singular filament composition. It is usually available in clear and smoke. Most brands are now heat resistant, but check to be sure.

WATER SOLUBLE: This thread will disappear with soaking or spritzing with water. Be sure to test your fabric first for water sensitivity. Water soluble threads are ideal for trapunto, basting, and reversible appliqué.

HEAT SHRINKING: This polyester thread creates a shirring effect, faux smocking, and puckers. Stitch and steam these threads and they will shrink up to 30 percent when used in the top and bobbin. Light weight fabrics are more easily affected.

FUSIBLE: The fusing feature of this white, floss-like thread can be used for utilility or decorative purposes. It can temporarily hold appliqué pieces or binding in place or can be stitched onto a background and then foiled with heat press foil sheets.

ELASTIC: Create a shirred effect without having to encase elastic. Hand wind elastic thread onto several bobbins, without stretching the elastic. Stitch in parallel rows with medium sewing thread on top and elastic in the bobbin to increase the shirred effect. Use a straight, open zigzag, or elastic stitch.

Thread Coloration

Terminology differs from one thread brand to the next, but these are some common terms related to thread coloration:

A variegated thread has random or regular color breaks along a band of values for one color. Some manufacturers call this an Ombre effect.

A multi-color thread is said to be space dyed. Space dyeing, also known as section dyeing, imparts different dye colors along the thread in regular or random intervals. The predictable color change can help direct your next quilt movement, while the random change can offer painterly effects or unexpected color surprises. Subtle or intense color changes can bleed into one another and create new colors.

The process of hand dyeing is commonly used by artisan thread producers and some larger companies. Some refer to this as hand overdyeing. Overdyed threads are dyed in one color and then dyed in random intervals with one or more additional colors. It is a good idea to get more than enough of your chosen thread to avoid disappointment.

Thread Weight

If you pick up three spools of thread from different manufacturers, each marked 40, you may find you have three threads that are not the same thickness at all. The confusion surrounding thread weight stems from the fact there is no commonly used measurement system for thread. The lower the weight number, the heavier or thicker the thread. While weight defines thread thickness, ply indicates the number of fibers twisted together to form the thread and give it strength. Three ply will generally be stronger than two ply. We encourage you to simply think about thread in 5 size groupings: Ultra fine, fine, medium, heavy, and ultra heavy.

Ultra fine: Threads listed with a weight of 70-100. This category consists mainly of polyester and silk threads.

Fine: Threads weighted between 50-60 mostly considered piecing and utility threads in cotton, cotton/poly, and polyester.

Medium: General sewing weight thread 35-40 weight, both utility and decorative. This includes most construction threads, cotton, rayon, and silk threads, as well as most metallic and specialty threads.

Heavy: 12-30 weight threads available in both cotton, rayon, wool blends, and silk. These heavy threads make a big impact in your stitch work.

Ultra Heavy: 5-8 weight crochet threads and heavy cord only to be used in the bobbin.

tip

- THE STASH: To begin building your thread stash we suggest you take a quick inventory of all the threads you currently have. Separate them by fiber, weight, and color. Make a list of the gaps in your stash and use this list to begin to fill in each category. Begin with a couple of neutral shades in each category to experiment and then expand your favorites to include all the colors you desire.

  Thread should not be stored in direct sunlight. If you live in a dry, dusty environment you will want to store your thread in an enclosed system. If you live in a more humid area you can probably store thread out in the open.

  Store your thread in the same way you approach your sewing. Debbie sorts thread by color and Liz sorts by fiber and size over color. Both methods are valid. We both store our thread in plastic drawer systems.

For more in-depth information on thread-related terms, visit www.ylicorp.com - Education; Thread of Truth brochure.

the threads

The chart below gives some examples of threads from different manufacturers in each category.

Ultra fine	Fine	Medium	Heavy	Ultra heavy
Presencia Finca Bobbin Lace 100 WonderFil, InvisaFil, Deco-Bob YLI Silk 100	Madeira FS 50 Metallic Marathon Underline (60) Presencia Cotton (50 & 60) Star Quilting (50) Superior So Fine, Bottom Line, Masterpiece Tire Silk (50)	Coats & Clark Dual Duty XP, Hand Quilting Gutermann Jeans (poly/cotton) Dekor (rayon) Kreinik ZTIP Madeira Polyneon, Ombre Marathon Cotton, Rayon Variegated Signature 100% Cotton Quilting Superior King Tut, Rainbows, Living Colors Weeks Over-dyed YLI Machine Quilting Most rayon and polyester decorative threads Most cotton threads	Aurifil Mako 12, Lana (wool) Coats & Clark Heavy, Button & Carpet Polyester Gutermann Heavy Silk, Heavy Polyester, Polyamide Metallic Kreinik Braid, Fashion Twist Linda Palaisy Soy Silk Madeira Lana (12), Glamour (12) Signature Size 20 Cotton, Pixelles Superior Perfect Quilter, Brytes Sue Spargo wool Valdani Pearl Cotton (12), Sewing Thread (30) Most wool/acrylic blend threads	Madeira Glamour (8) Presencia El Molino & Finca Pearl (5 & 8) Superior, Razzle Dazzle Valdani Pearl (5 & 8) Pearl cottons

This chart represents the authors' determination of thread weight by category based on manufacturer's description and samples available.

tip

CROSS WOUND AND STACKED SPOOLS

A closer look at your threads will reveal stacked threads are wound around the spool in a parallel fashion while cross wound threads criss-cross each other in overlapping X's around the spool. Stacked spools unwind best from a vertical, upright position, while cross wound spools perform best on a horizontal spool pin, secured with a spool cap. If your thread misbehaves, try changing its position.

The Passport

A passport officially sanctions travel abroad to both pedestrian and exotic locales. The visas and souvenirs you gather document your journey. Your stitched passport similarly documents your stitch travels. In it, you can record all of your stitching adventures and maybe even a few detours and daring misadventures!

Why create a passport? The small size of the passport page allows for fearless experimentation and play without the pressure of wasting time and materials on a huge project. The beauty of the passport is the fact you can always add to it. Whenever you want to try a new technique, you can document it with a new page.

This book is designed to be worked from front to back. Like a traditional stitch sampler, skills are learned and then built upon. We recommend you do each exercise. A few may seem tedious but each teaches very important skills or explores a fundamental concept. We also encourage you to try each technique, even the ones you may find intimidating or simply not to your taste. Our students are often amazed when they tackle a technique or discover a new avenue of stitch exploration they wouldn't have normally chosen.

Sometimes it is more fun to travel in a group. Consider forming a stitch adventure club with friends. Get together weekly or monthly to experiment with techniques and share ideas. A group can encourage you to continue stitching and learning when daily duties attempt to derail your creative adventures.

making your passport pages

Ready

A ***passport page*** consists of fabric, batting, and stabilizer. Each page is constructed separately and then stitched back to back to assemble into a book. We chose 5"x 7" for our passport but choose any size or shape you are comfortable with. You may have to make some minor adjustments to our directions if you choose a different size. We pre-cut fabric, batting, and stabilizer so it is ready to go when we are. We found it easiest to cut our supplies about 1/4" larger than our desired finished size to allow for shrinkage.

tip

Portfolio pages are used for techniques that require more space than the typical Passport page. Portfolio pages are made with two 12" squares of fabric with a layer of low loft poly-cotton batting between them.

Set

Choose your binding technique, grommets or eyelets set directly into the page or tabs to bind your pages. Refer to instructions on page 12 for both binding techniques. Read through the directions before beginning to stitch. Your binding choice may affect how you create your pages.

Stitch

To assemble the separate pages into a book, choose two pages that will be placed back to back. Pin them together and stitch around the perimeter with a straight or zigzag stitch to secure. An even feed foot is helpful to keep the page edges lined up. Choose an edge finish, referring to the techniques on page 36.

tip

PASSPORT COVER
Save the best for last! After you play, create a cover for your Passport using your favorite techniques. Add the word 'Passport' or come up with a fun title that makes your book truly yours.

Bind

Set grommets, stitch buttonholes, or attach tabs to your pages. Use binder rings to clip the pages together. Embellish the rings with ribbon, yarns, and torn strips of fabric.

Grommets, Eyelets, and Tabs:

A little pre-planning is necessary to ensure you have enough room for binding holes when using the grommet or eyelets set directly into the page. The left hand page will need room for the binding holes on the right side of the page. The right hand page will need room on the left side of the page. Keep the area around the holes free of stitching whenever possible. Create a template to mark the binding holes ahead of time. Use plastic template material, sturdy cardboard, or heavy duty stabilizer cut to the size of your page. Mark placement holes 1-1/4" from the top and bottom edges and 1/2" in from the side of the template. Use a hole punch to make a circle for marking the grommet placement on your fabric page.

When your pages are ready to be bound together, use your template to mark the hole placement on each page. Create a hole through all the page layers using very sharp small scissors or the Crop-A-Dile™ tool. Refer to the directions of your specific grommet setting tool for instructions on setting grommets and eyelets. As an alternative, stitch small buttonholes directly in your pages.

making your passport pages

Tab Binding:

The tab binding is a bit more labor intensive, but is sturdier and allows for full use of the space on each page. The individual tabs may be any size you like, but be sure to keep the placement for the hole or buttonhole consistent on each tab.

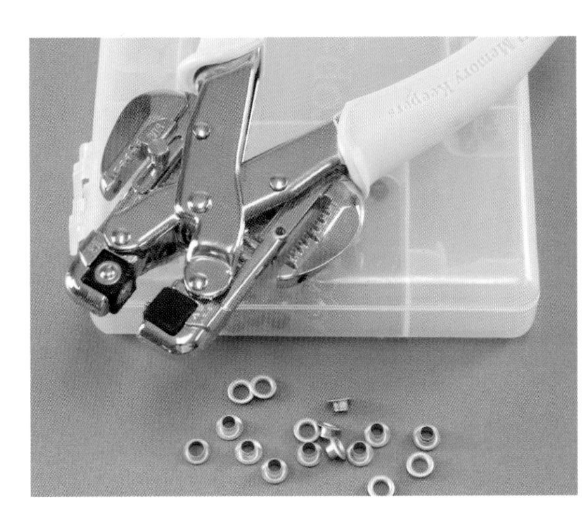

Grommets & Eyelets: The main difference between grommets and eyelets is the size of the flange in relationship to the hole. Grommets have a larger flange and offer more security for binding your Passport pages. We use a 3/16" grommet or eyelet. Grommet setting kits are very inexpensive or you can choose a Crop-A-Dile™ tool. It is designed for eyelets and is great for punching the holes through the fabric as well as setting the eyelet.

Steps

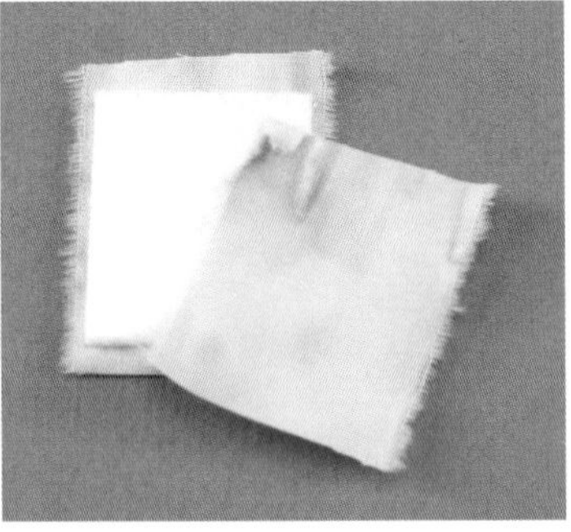

1. For the tabs on a Passport page, cut two 1-1/4" x 1" pieces of heavy weight stabilizer. A stabilizer with a fusible surface makes this technique faster or you can add a layer of fusible web to your stabilizer. Cut two pieces of fabric slightly larger than 1-1/4" x 1". Fuse the fabric onto both sides of the stabilizer. Stitch around edges if desired.

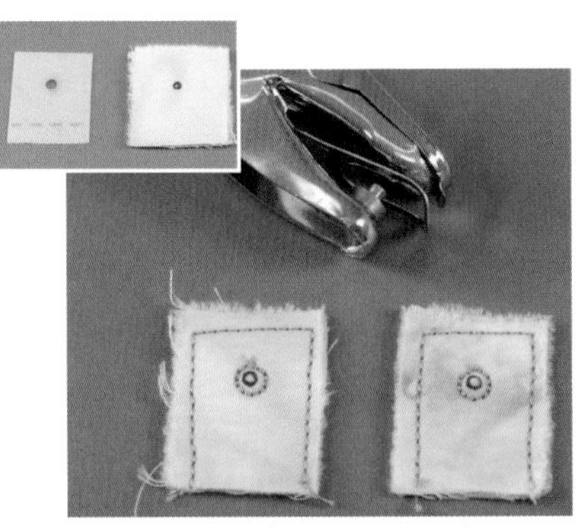

2. Use a marker or pencil to mark the hole or buttonhole placement on your tab. Use the template provided or create your own. Stitch an eyelet or buttonhole. Cut out the opening with small sharp scissors. If you use a stitched eyelet you can use the Crop-A-Dile™ tool or Fiskars® craft punch.

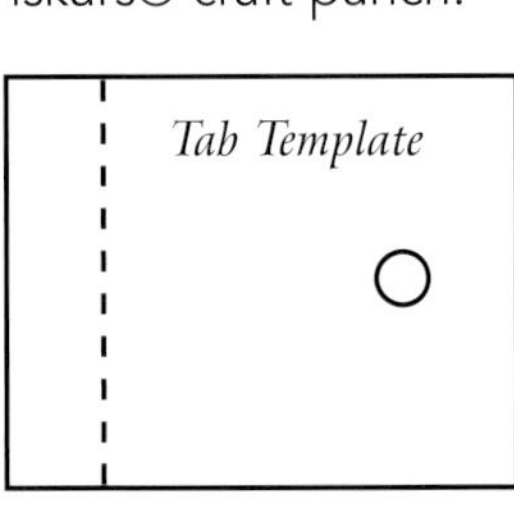

3. Insert the tabs between two pages and stitch in place. Add a decorative edge finish. You can also stitch the tab onto the surface of a two page sandwich. Pre-plan this option and leave the area free of stitches. Insert binder rings in the eyelet or buttonholes.

packing for your journey

Preparing for a grand adventure may seem a daunting task. What do you need to take, what needs to be bought, what can you make do with or without? Use our guide to help you pack and soon you will be on your way to exploring the mysterious and exotic world of thread.

SEWING MACHINE:

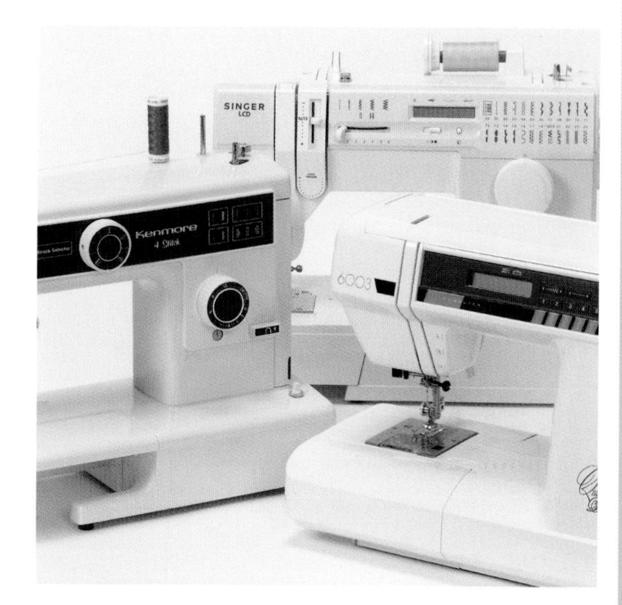

A basic sewing machine is all that is needed for this book. While a large assortment of programmed stitches is fun to play with, magic can be made with simple straight and zigzag stitches. The ability to drop or cover the feed dogs will make free motion stitching smoother and easier. Clean and oil your bobbin case frequently. Refer to your machine manual for specific directions for your machine.

Your sewing machine is an investment and the two minutes it takes to clean and oil it will pay off with a longer life and fewer repairs.

Always use the bobbin designed for your sewing machine. "Close enough" isn't close enough! Generic or off brand bobbins can cause poor stitching or costly damage.

SPECIALTY SEWING MACHINE FEET:

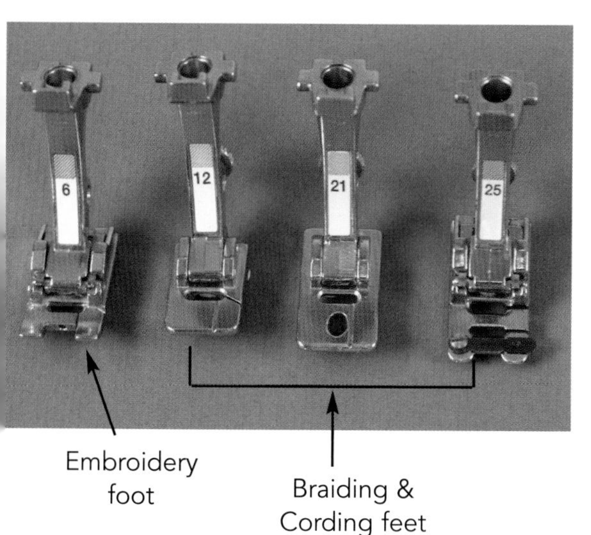

Embroidery or appliqué foot: These feet have an open front to allow a better view of the stitching area. A groove underneath accommodates the build up of thread when doing dense stitching. Fine yarns or ultra heavy threads can be threaded through the hole in the front of the foot for couching.

Beading, Braiding or Cording feet: Use these feet to allow space for yarns and fibers to be guided under the foot for couching. There are a variety of foot options to accommodate different types of fibers.

Walking Foot: This foot is designed to evenly feed fabric under the presser foot. This is an invaluable foot for machine quilting.

Darning or free motion foot: This foot is used with the feed dogs dropped. It allows the fabric to be moved easily in any direction for free motion stitching.

NEEDLE SIZES & USES:

Universal:
Available in sizes 60-120, the point is slightly rounded for use on knits, but they are still sharp enough for some woven fabrics. They work better with loosely woven fabrics.

Microtex/sharps:
Available in sizes 60-110, the point is sharp and features a narrow shaft that works well with tightly woven fabrics. They are good for quilting, heirloom stitching, topstitching, and edge stitching.

Quilting:
Available in sizes 75-90, this needle features a slightly rounded, tapered point and is designed to stitch through thick layers and intersecting seams. Good for quilt piecing and general quilting.

Embroidery:
Available in sizes 75-90, these needles have a light ball point and a deeper scarf on the back to help with loop formation with fine threads. The larger eye reduces friction and allows you to use specialty threads with ease. They are designed to stack the stitches close together without damaging the threads next to them. Embroidery needles were the first needle designed to stitch in many directions.

Topstitch:
Available in sizes 80-100, these needles are designed to go through layers of fabric or heavy fabrics. They have a larger shaft to reduce flex and a sharp point to aid in obtaining a straight line. They also have a very long eye and deeper groove for heavier threads or to accept two threads at one time.

Metallic:
Available in sizes 80-90, this needle was designed specifically for use with metallic thread. It has a larger, more elongated eye than an embroidery needle and its fine shaft and sharp point reduces friction on the thread. It also has a deeper scarf and a deeper groove. The metal in the thread can wear a groove in the needle so it generally has to be changed more frequently.

- Use an empty spice jar with a hole punched in the lid to safely dispose of old needles.
- Just for fun pick up twin and triple needles or some wing/hemstitch needles.

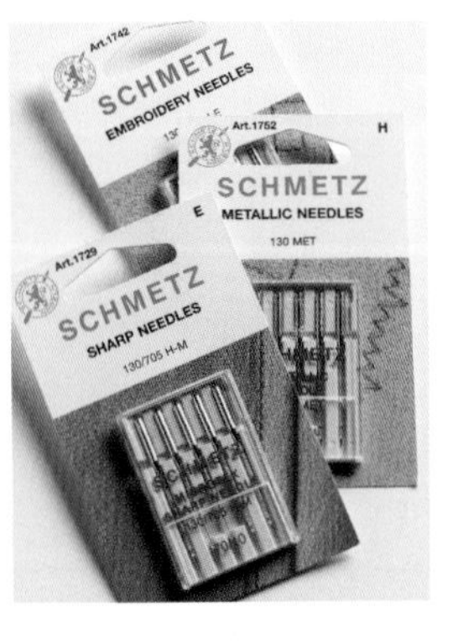

Needles:
Today's stitch enthusiasts have a wonderful array of needles to suit every fabric and thread. We are often unsure of which needle to choose and fall back to using a 'universal' needle, which is rarely the best choice. Choose high quality needles designed to work with your specific machine. Needle sizing is opposite of thread sizing. The larger the number, the larger the needle. Generally you should use the smallest needle that accommodates your thread size. Change your needles often, at least with every project if not more often. Keep a stash of your old needles for sewing through paper, metal, and mica and then throw them out. Dull needles will cause excess friction on your thread, invite skipped stitches, and snag delicate fabrics.

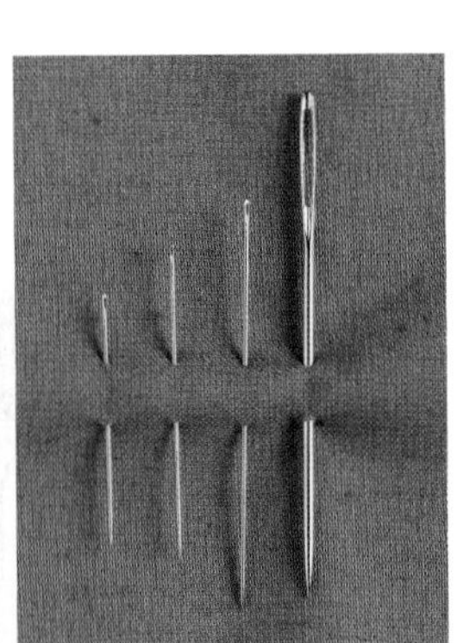

Hand Sewing Needles:
Hand sewing needles are sized opposite of machine needles, the higher the number, the smaller the needle. Have an assortment of hand sewing and embroidery needles on hand, as well as a few larger eye chenille needles for free form embroidery. Needle threaders are handy tools if threading the needle is difficult. Beading needles are very thin with a small eye to easily go through the hole in a bead multiple times. They are available in different sizes from the thinnest size 16 to a thicker size 10. Beading needles generally come in short and long lengths. Many beaders use milliners needles.

packing for your journey

FABRIC:

The Passport is a great way to use fabric bits already in your stash. For your Passport page bases you will want fairly plain mottled fabrics that will let your thread work shine. Use larger prints and patterns for accents, but don't limit yourself to just cotton. There is a wonderful world of specialty fabrics just waiting to be discovered and used. We like to rummage in the remnants section of any fabric store and are always on the lookout for garments and vintage linens. Try adding Lutradur®, felt, paper, thin metal sheet and foil, Angelina® fibers, silk paper, and anything else you can stitch through.

HOOPS:

Embroidery hoops are useful for holding your fabric taut while adding stitching and will help prevent hand fatigue when thread painting. Spring hoops and screw hoops are useful for machine thread work. Spring embroidery hoops are usually plastic rings with a metal inner ring. The inner ring is squeezed to remove it from the hoop and released to hold the inserted fabric taut. Use this type of hoop for small projects where a lot of additional stitching is not desired. Screw hoops are made of wood or plastic and have a small brass screw mechanism to tighten and release the hoop. Purchase a hoop the right size for your machine. The radius cannot be larger than the distance between your needle and the inside of the machine bed.

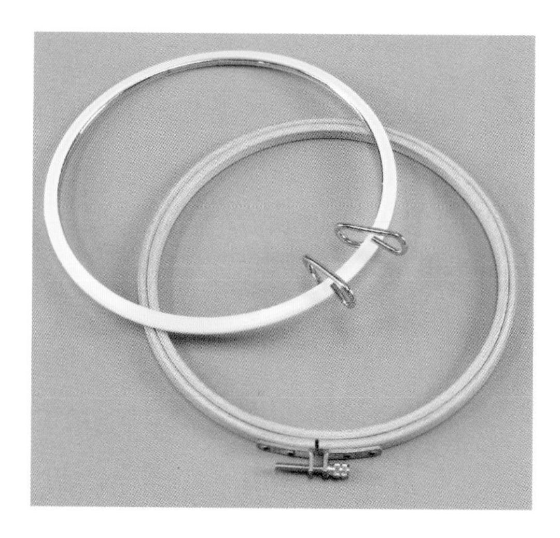

THREAD STANDS:

A separate thread stand can be helpful when using large spools of thread. It also gives some unruly specialty threads, particularly metallics, extra space to unspool without twisting and kinking. If you don't have a thread stand, try standing your spool in a cup or jar beside your machine. Tape a safety pin on your machine as an extra thread guide if needed. Always take care to insure the thread cannot become tangled in the machine's fly wheel.

tip

- If thread puddles off from the spool or kinks into loops before entering the thread path, a thread net or vinyl wrap can control the wayward fiber. These provide just enough drag on the thread to keep it from flowing too quickly from the spool. Vinyl wraps are also a great tool for storing thread.

BOBBINS:

Bobbins are inexpensive and you can't have too many. There is nothing more frustrating than having to unspool an entire bobbin because it isn't the right thread for your project. We use rubber bobbin rings to corral and organize our bobbins.

BOBBIN CASE:

If you are hesitant to adjust your bobbin case to use heavy threads, the investment in an extra case will bring you peace of mind. The screw adjustment on your bobbin case follows the righty tighty, lefty loosey rule. Work over a table or small box when loosening your tension screw in case you go too far and it pops out. Most bobbin cases will only need small adjustments to give you the proper tension for thicker threads. To understand small adjustments visualize a watch face and the distance for five minutes. You can count how much you need to turn the screw to loosen it and then return it to normal. For some threads you can simply turn the screw five minutes, others may need 20 minutes. Make note of the specific alignment of the bobbin case screw before making any adjustments in order to return it to the exact alignment. With a rotary hook, front loading bobbin case, you may find it easier to rely on 'pull' feel to adjust your case. Thread should pull out of the case housing smoothly and at a steady rate. You should not have to tug to pull it out nor should it come flying out with a light touch. A little practice with this method will give you confidence to adjust your bobbin case. Don't forget that when using finer threads you may need to tighten your tension to achieve a perfect stitch.

Adjusting the tension on a horizontal, drop-in bobbin case is similar to a rotary hook, front-loading bobbin. Follow the righty tighty, lefty loosey rule and turn in small five minute increments. There may be two screws visible on the horizontal case. Look for the one connected to the tension spring within the case. Replace the bobbin in the case and draw the thread through the usual thread path. Cup the case in your hand and draw the thread through the bobbin spring. Only a slight amount of pressure need be applied to draw the thread through the case. Eventually you will get a 'feel' for the tension your bobbin's thread is under. Be careful not to let bits of thread build up in the case's thread path.

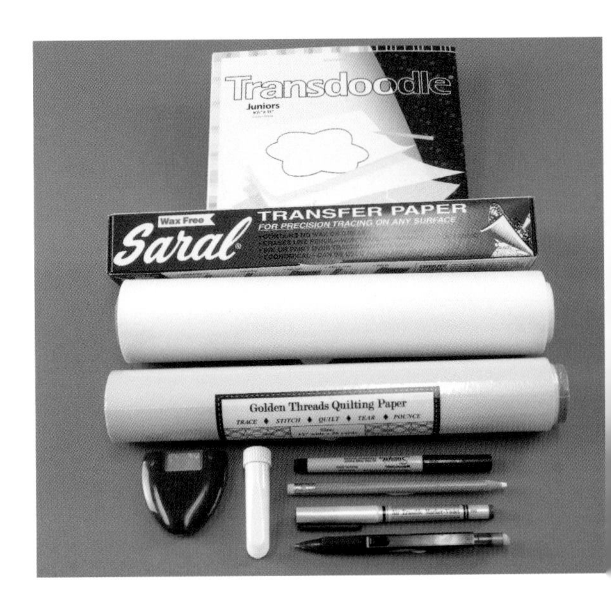

MARKING TOOLS:

Marking tools range from pencils, chalk wheels, air and water disappearing markers, hera markers, and permanent markers to transfer paper such as Transdoodle® or Saral®. Have a variety on hand for different marking jobs.

Light weight paper such as drafting velum, tissue, or Golden Threads paper are great for marking patterns and lines without actually marking on your fabric.

packing for your journey

COLOR WHEEL:
The color wheel is one of our favorite and least expensive tools. Basic information is generally printed on a color wheel and is enough to get you started.

Playing with the warmth of color in thread painting can establish a light source or convey the mood of our stitchery. Think of red-violet and yellow-green as forming a horizon line across the color wheel. The reds-yellows-oranges above this divide are considered the sunny, warm colors. Those below this line are the cooler, under-the-sea greens-blues-purples.

A color's *value* is its relative lightness or darkness along a scale of gray that travels between white and black. By using contrasting values of colors in thread painting, we can establish shape, contour, light source, and shadow.

When building a thread stash, think of color, weight, and value. Subtle changes in the value of analogous colors are particularly important for stitching portraits.

Pump up the excitement of your stitching with complementary (opposite) thread color or the split complement (those to either side of that opposite) to the predominant color of your fabric. For a more subdued effect stick to analogous (side by side) or monochromatic (tints and shades of the same hue/color) color schemes.

EMBELLISHMENTS & OTHER PLEASURES:
After fabric, the things that add excitement to the journey are embellishments. An assortment of yarns, ribbons, lace, and fancy trims are fun to have, as well as beads, sequins, buttons, found objects, feathers, rubber stamps, and ink. There are no limits to the cool stuff you can add.

FROM THE KITCHEN:
Use freezer paper for making temporary templates. Parchment paper can be used as a pressing sheet and a cookie sheet back is perfect when using heat tools on fabric. Bamboo skewers can be used to hold yarns and small fibers in place under the sewing needle.

CUTTING TOOLS:
Rotary cutters, rulers, and cutting mats are optional but they make measuring and cutting fabric precise and easy. We recommend at least one small and one large pair of sharp scissors.

BATTING:
The Passport pages are a great place to use leftover batting scraps. In most cases the batting material is not important.

getting started

Fusible webs and stabilizers have changed the way we quilt forever. There are two main aspects of a fusible product that need to be considered: the strength of the bond and how it affects the drape of your fabric. Stabilizers are designed to stabilize the fabric before you stitch on it. Keep in mind the more stitching you plan on doing the sturdier the stabilizer you need. Making a sample page will help you understand first hand both of these factors. Refer to Resources on page 175 for a list of commonly used fusible webs.

FUSIBLE WEBS:

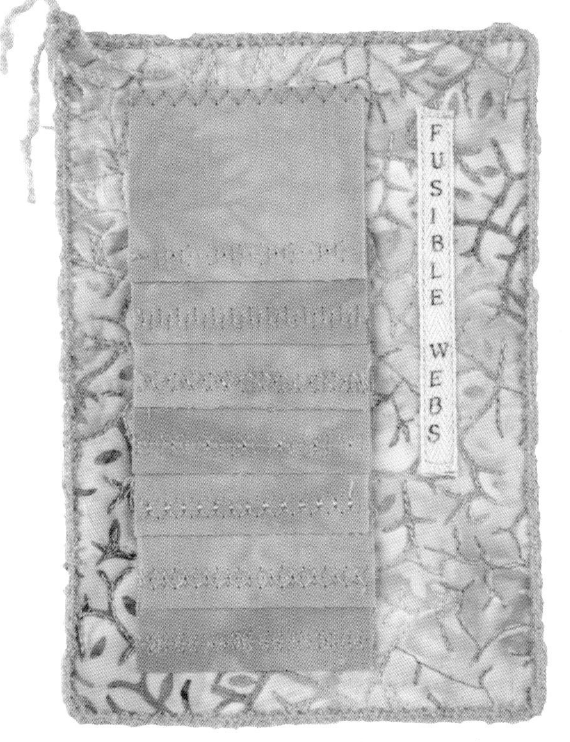

Edge finish: couched yarn with zigzag stitch

Fusible webs are also known as dry adhesives. They are a fabric made of strands of glue and when heated will activate to fuse two flat objects together. There are many different brands and varieties but all use an iron to activate the glue. Most of the available products can be used interchangeably but some are heavier than others and will change the feel of your fabric. The light weight fusibles do not come with a backing paper so a non-stick craft sheet, like the Goddess Sheet™, is helpful to avoid getting the product on your iron. Light weight fusibles are very helpful when using light weight fabrics, Angelina®, and paper or to reduce stiffness when using multiple layers of fabric. Medium weight fusible webs come with a release paper on one or both sides. Some are tacky and will temporarily adhere to your fabric until they are bonded. Refer to the specific directions on the product you are using and follow their recommendations. Never place the iron directly on the fusible web because it will melt onto the iron sole plate. If this happens use an iron cleaner or rub the sole plate on a scrap of fabric.

STABILIZERS:

Stabilizers are often overlooked as a basic stitching tool but they help control fabric distortion and support, or even replace, fabrics when adding lots of stitching. Stabilizers come in many varieties. The more thread you want to add, the heavier the stabilizer should be. Fortunately, they are generally inexpensive so you can try many different types without breaking the bank.

Tear Away Stabilizers:
These stabilizers are meant to be used on the back of a fabric when stitching and then torn away. They may be left in if desired.

Cut Away Stabilizers:
Cut away stabilizers are applied to the back of the fabric and are intended to be left in place after stitching or embellishment. Excess stabilizer is cut away from the embellished area.

Interfacings:
Generally designed for garment construction, interfacings are great stabilizers and are often fusible on one side. They are especially useful for stabilizing unruly fabrics such as silks and satins. Many brands and weights are available.

Heavy Weight Stabilizers:
Heavy weight stabilizers are very helpful when a stiff base for your project is needed. They are often used when creating fabric postcards, vessels, and books. When creating a densely stitched piece, these stabilizers can replace fabric for a stitching base. They vary from thin heavy weight to a thicker padded heavy weight.

Heat Reactive:
Heat reactive stabilizers crumble and turn to dust when heat is applied. They are a wonderful addition to your toolbox for projects where you don't want to subject specialty fabrics to water or the stress of tearing away other stabilizers.

Water Soluble Stabilizers:
The most common use of water soluble stabilizers is in the creation of thread lace, thread structures, and appliqués. The stabilizer washes completely or partially away with water. There are different weights and types of water soluble stabilizers. Some stabilizers look like a soft towel and are opaque. We prefer the transparent water soluble stabilizers that allow you to easily trace a pattern on the stabilizer. The lighter weight stabilizers are ideal for use in a double layer when you are creating a piece that will incorporate a delicate fabric, netting, or fibers. The heavy weight stabilizers can often be used without a hoop as a base for heavy stitching. Some water soluble stabilizers come with an adhesive on the product. The adhesive does not gum up the needle and washes away completely. The adhesive allows exact placement of fibers, fabrics, and threads when creating a design to embellish with stitching.

Refer to Resources on page 175 for a list of commonly used stabilizers.

beginning your journey

Follow the steps below to make two Passport pages, one using a variety of fusible webs and one using a variety of stabilizers.

Machine set up:
Normal

Tension:
Balanced

Needle:
Embroidery 90/14 or sharp 80/12

Top thread:
Any decorative thread

Bobbin:
Light or medium weight thread

1 Cut 2" x 2-1/2" pieces of fabric for the sample tabs.

2 Cut 2" x 2-1/2" pieces of assorted fusible webs and stabilizers.

3 Place a Teflon pressing sheet or a piece of parchment or light weight paper on the ironing surface to protect it and the iron from the glues. Following manufacturer's directions, fuse two pieces of fabric together with each different type of fusible web and fusible stabilizer. You may want to place a second piece of paper between the iron and fabric.

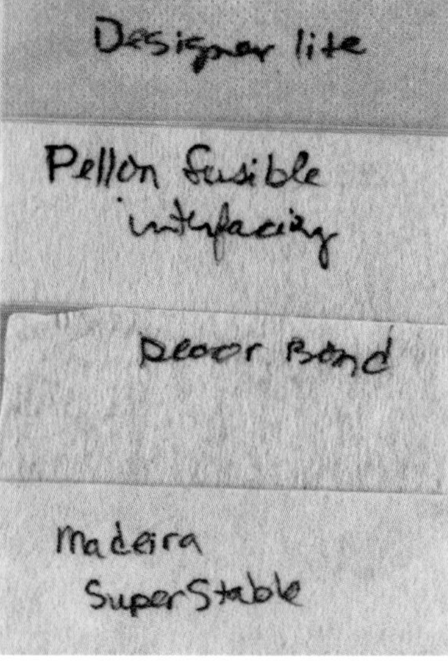

4 Write the name of the fusible web or stabilizer product on the back of each piece.

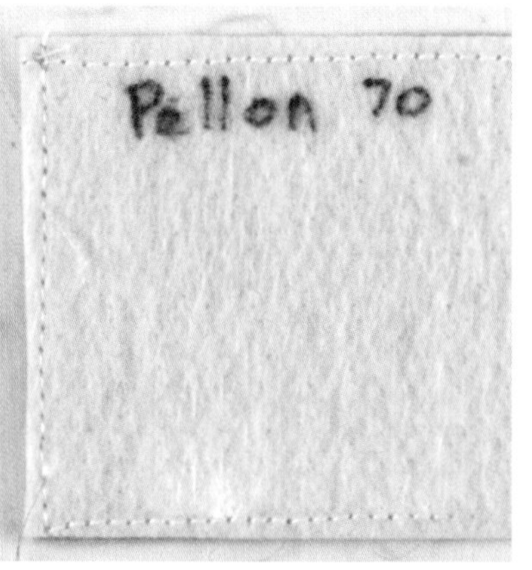

5 For the sew-in stabilizers, lay the fabric piece on the stabilizers. Pin and stitch around the edge.

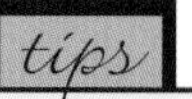

tips

- Store stabilizers in a sealed plastic container or bag for longer shelf life.
- Before rinsing the stabilizer away, trim away as much excess stabilizer as possible. The scraps can be misted with water and patched together to form a larger piece or dissolved in water and sprayed on a fabric to stabilize it for stitching. If you develop a tear or hole while you are stitching, make a patch by lightly dampening the area around the hole and apply a small piece of stabilizer over the area. It will adhere and you can resume stitching.
- Two layers of stabilizer may be sealed together with a warm iron.

6 Sew a decorative stitch on the bottom edge of each sample if desired. This step demonstrates the stabilizing effect of each fusible web and stabilizer. Layer a page fabric, batting, and tear away stabilizer sheet together. If desired, quilt the base fabric.

Place the fusible samples on the page base and attach with a straight, zigzag or decorative stitch.
Repeat to make a stabilizer Passport page.

beginning your journey

Each Passport page is a stitching opportunity and some of those pages might be quilted. Whether large or small, for a bed or a wall, we recommend these steps for quilting a quilt —Plan, Mark, Anchor, and Baste.

Plan

- Thread can play different roles in your final quilt. Color, thread weight, and special effects are all important considerations to achieving a contemporary, folksy, reproduction or other style. Fine or ultra fine threads that sink into the quilt become nearly invisible offering more texture than visual effects. Medium weight threads can show stitching designs and blend or stand out from the fabrics. The heavier threads offer a bolder stitch statement, and ultra heavy threads can be bobbin stitched as part of the quilting mix.
- Do a rough sketch of your quilt pattern and photocopy it. Then draw your stitch pattern on top of the quilt design. Try a few different versions and variations, since first ideas are not always the best ones. Use plastic or tissue paper overlays or mark a design with removable marking tools or painter's masking tape. Refer to page 16 for information on marking tools. Finally, do some trial stitch outs. Assemble a variety of potential thread candidates and play with them. See how they look on your quilt fabric. You might be surprised to see how things come together, and a whole new plan might emerge.

tip

- To determine if your quilt and thread will stand up to washing and drying, piece a few scraps of the quilt's fabric together and stitch with your proposed threads. Measure the sample, toss it in the washer and dryer to see if it shrinks. If it does, make adjustments. Plan now before the quilt is finished.

Mark

- Sometimes it is best to mark the quilting design on the quilt top BEFORE it is basted, as the fabric stays flatter. If you are marking before basting, use a method that will withstand rubbing, rolling, and possibly ironing the quilt. Other quilts can be completely marked after basting, while some need to be marked one section at a time, either because of the marking method or the quilting is happening in stages or layers.
- Passive marking methods do not involve actually putting a mark on the quilt. These include pins, tape, hera markers, and crease lines. Tacking a quilt only requires marking the tacking spots with pins or tape. Repetitive straight lines might best be marked with regularly spaced tapelines. Tissue paper is also a passive method, as the design is marked on the tissue and stitched through to the quilt. The tissue is then easily removed. Tissue is versatile for marking a quilt top, reverse bobbin work, and thread painting.
- Active marking methods require making actual marks on the quilt. What you use will depend on the type of fabric you are marking and the design you are transferring or drawing. Methods include a variety of chalks, disappearing or wash-away markers, and transfer paper. A light touch with a fine pencil, regular lead or colored, is also acceptable. If you are stitching over the lines, they will be nearly invisible. Antique quilts are often prized for the occasional remnants of the quilter's handmade markings.

Basting

- Basting is the temporary binding together of layers in preparation for final sewing. Because of its temporary nature, you may be tempted to skip it, but quilts come together faster and flatter with adequate basting. We stress "adequate" because just adding a few pins is not enough. You will find yourself 'plowing' a pile of excess fabric across your quilt.
- Small pieces can often be basted with straight or safety pins or spray basted with temporary spray adhesive. For larger projects, use safety pins, a basting gun (fine plastic tacks) or simply hand stitch with large running stitches in a contrasting thread that will be easily seen, stitched over, and pulled out when the job is done. You can also baste with water-soluble type threads that will be gone from the quilt after its first washing.

Anchor

- Anchoring may be needed when using embellishing threads in the quilting plan. For example, if we use a metallic thread on a quilt for its embellishing properties, we are often sacrificing strength. Anchoring stitches are most often done with light weight, but strong polyester or cotton threads that blend into the background or are hidden within the ditch. They do the work of holding the quilt together. Then, flashy embellishment threads capture the spotlight without you worrying about the quilt falling apart.
- Anchor stitch by stitching from the center of the quilt to the outer perimeter, gradually spreading all of the quilting stitches out across the quilt and not quilting one area too heavily in regard to the remainder of the quilt.

stitch dictionary

FREE MOTION STITCHING:
In free motion stitching the sewer moves the fabric under the needle instead of letting the machine's feed dogs do the work. Disengage or drop the feed dogs on your machine and use a darning or free motion foot.

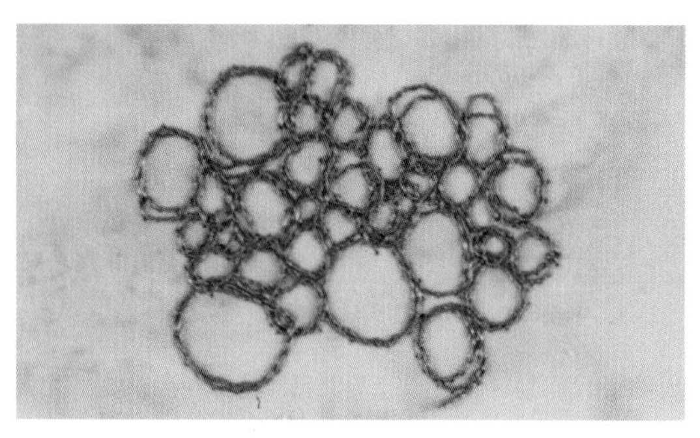

GARNET STITCH:
Also called granite and circle stitch, this free motion stitch is made when the fabric is moved in circles to create a pebble pattern. The pattern is often stitched twice for emphasis.

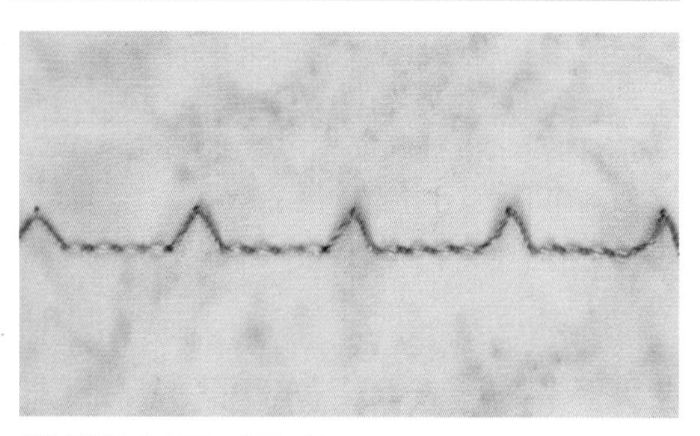

INVISIBLE HEM STITCH:
This utilitarian stitch is used for hemming garments and drapes. The invisible hem stitch uses a tailor tack foot. It can also be made with a straight stitch foot for appliqué or a couching foot for couching thick yarns.

STIPPLE STITCH:
A stipple stitch is any quilting stitch pattern created closely together (1/8" - 1/16" apart) to add texture to the fabric surface and fill in areas. Derived from the drawing term 'stipple', this describes the act of marking and shading with dots or tiny lines.

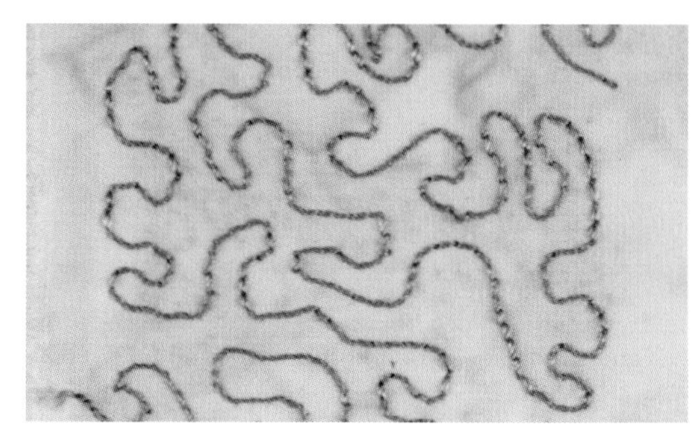

MEANDER STITCH:
This stitch is also known as vermicelli stitching and is often mistakenly called stipple. The stitch is done with the machine set up for free motion stitching and is most often used as a quilting stitch. The stitching meanders around the surface in curves that may approach each other but do not cross.

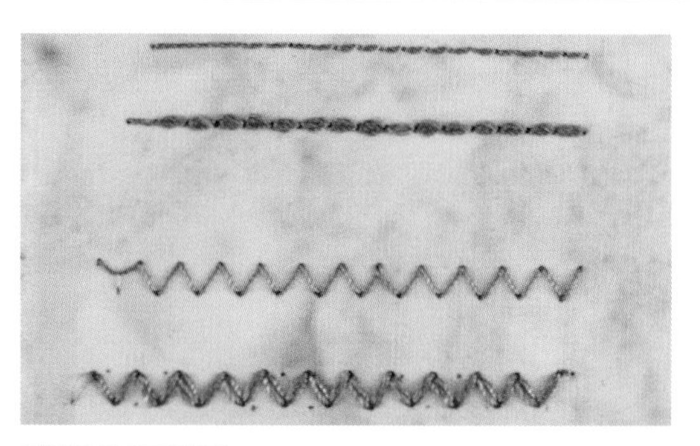

TRIPLE STITCH:
This stitch is frequently indicated on your sewing machine by three straight stitch lines next to each other. This stitch is often recommended for use with heavy fabrics such as denim but has decorative uses as well. Setting the width of the stitch wide will create a triple zigzag stitch.

PROGRAMMED DECORATIVE STITCH:
A patterned stitch that is pre-programmed in your machine is referred to as a programmed decorative stitch. These stitches can be stitched as is or adjustments may be made to their length, width, and density. While the stitch is designed to be used with the feed dogs up, it is possible to lower the feed dogs and use the stitch in a free motion manner to achieve unique textures and marks.

SATIN STITCH:
Traditionally the satin stitch is created using a zigzag stitch set at '0' length. It can be made with a narrow or wide zigzag. Many programmed decorative stitches are worked solidly and may be referred to as a satin stitch to describe the density of stitches.

programmed stitch guide

This may seem like a tedious exercise but you will be amazed how helpful it is to actually stitch out your programmed decorative stitches. Many look nothing like the small diagram.

Machine set up:
Normal

Needle:
Embroidery 90/14

Tension:
Balanced

Top thread:
Any cotton, polyester, silk, or rayon

Bobbin:
Light weight or matching

1. Cut a strip, 4" by the full width of fabric from a solid black and a solid white fabric.

2. Sew the two fabric pieces together along the long edge, press seam open or to one side. Back with a piece of tear away stabilizer.

3. Choose a medium value thread color. A medium value will help show if the stitch looks better with high or low contrast.

4. Begin with stitch number 1 on your machine and sew across both the white and black fabric.

5. Note the stitch number with a permanent marker and repeat until you have sewn all of your programmed stitches.

the techniques

the threads

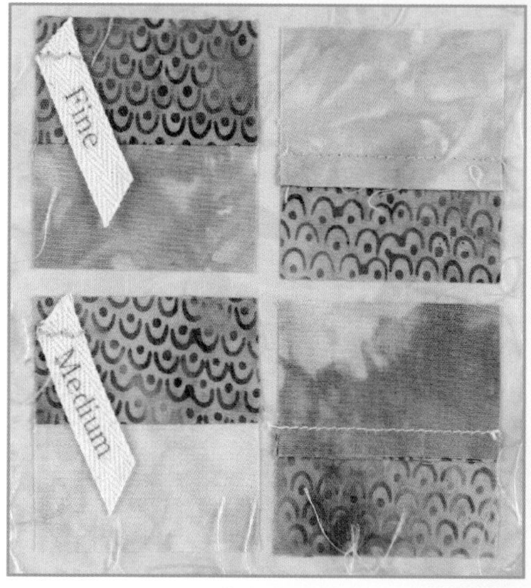

piecing

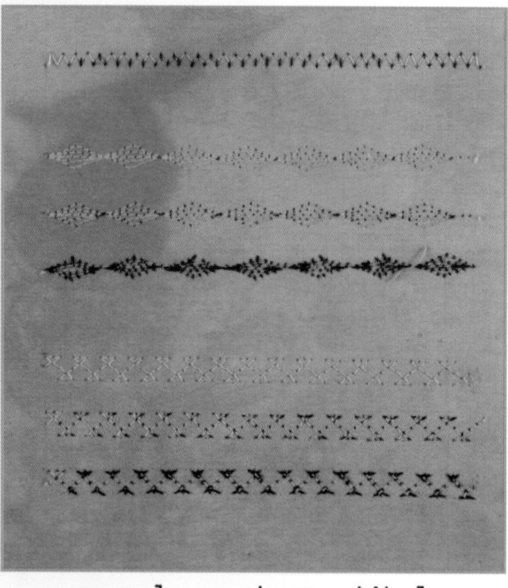
mock cretan stitch

whip & feather stitches

moss stitch

edge finishes

Nuts & Bolts

Learning the tools of the trade can seem a bit boring but it lays the foundation to further exploration and play. Our first set of exercises will help you master all the tools you need to boldly throw open wide the door of thread exploration. The individual exercises in this section can be worked in any order. Start with tension if you like, skip around, and have fun!

the threads

Since we are building a collection of all of the amazing things we can do with thread, why not start by stitching a sampler of our many options? It's a great way to find out how various weights and types of threads will work in your machine.

Edge finish: open zigzag stitch around the page followed by a hand-dyed ribbon tacked in place with a medium satin stitch.

getting started

Thread samplers can take any form you desire, from simple lines of stitching to ornate appliquéd elements. Choose a patch of free motion stitching or fill shapes with straight or decorative stitches. Pick a shape, fill it with stitches, and repeat.

1 Make a plastic or cardboard template of the sampler shape you would like to use. Be sure it will fit multiple times on a page. Back the fabric you are using with fusible web. Trace the template onto the paper backing of the fusible web. Cut out the shapes and remove the paper backing.

2 Arrange the shapes on the fabric page and fuse in place. Layer the fused fabric page, batting, and stabilizer. Stitch a variety of stitches on each shape using one weight of thread per shape. Change needles as dictated by the different thread sizes. The ultra heavy thread is stitched from the bobbin. Refer to Sewing with Ultra Heavy Threads on page 58. Stitch the same pattern on each template to have a fair comparison of the threads. Add words to your page if desired. Refer to Rubber Stamping on page 170.

Quick sampler: Thread sampler pages need not be time consuming. This quick stitch option will have you traveling to new pages in no time.

1 Trace circles using a spool end on fusible-backed fabric. Limit the sampler to the five basic categories.

2 Cut out and fuse the circles in place. Free motion a small meander stitch and splash of straight stitching in each thread.

3 Label each circle with the thread weight used. Use iron-on transfer sheets pressed onto ribbon or refer to Printing on Ribbon on page 169.

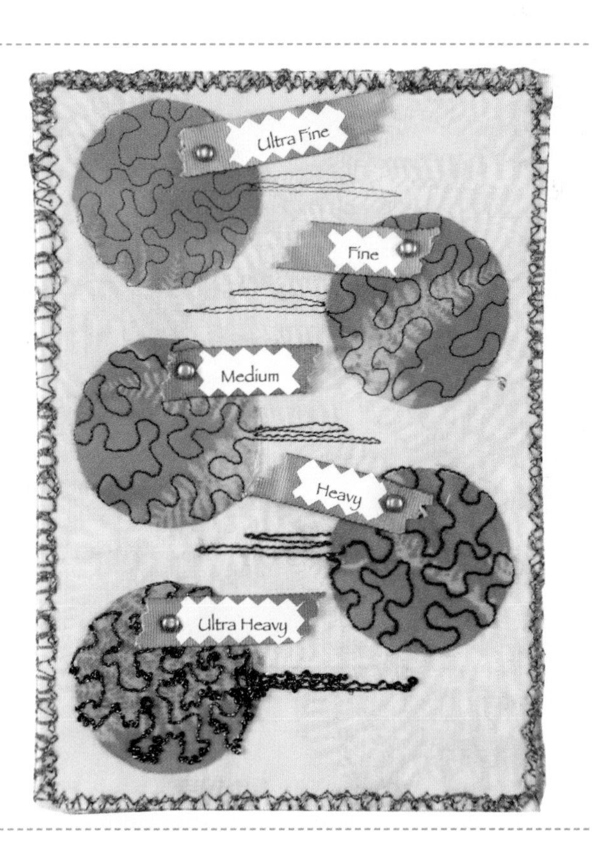

understanding tension

Often, the most misunderstood and feared aspects of a sewing machine are the tension adjustments. Understanding tension and how to make adjustments to your machine are crucial to fully exploring the numerous possibilities in all the wonderful specialty threads available today.

Edge finish: ribbon attached with a zigzag stitch

getting started

The Passport tension grid page will help you understand the specific tension settings for various specialty threads on your machine. We use a zigzag stitch to check tension because at the points of the zigs and zags any bobbin thread coming to the top will be easily seen.

Machine set up:
Normal and zigzag

Tension:
Variable

Needle:
Embroidery 90/14 or sharp 80/12

Top thread:
Various decorative threads

Bobbin:
Ultra fine or fine polyester in a contrasting color to the top threads

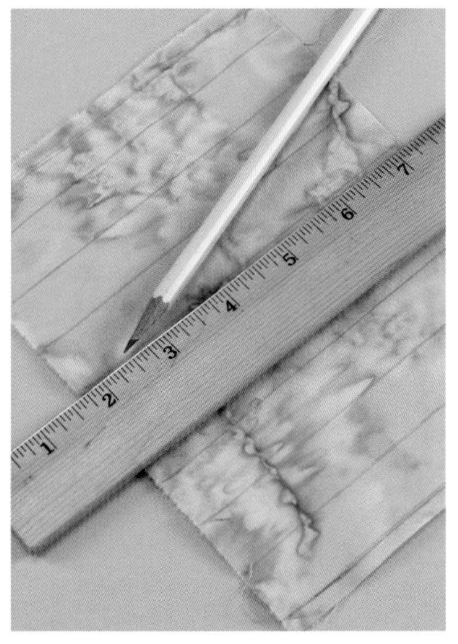

1 Draw horizontal grid lines on your fabric 3/4" apart. Stitch across the drawn lines if desired.

2 On the left side, write the numbers 0-8. These are the tension settings you will test. It is not necessary to go to a 9 or 10 tension setting because the fabric will usually begin drawing up and puckering when the tension is this high. If your machine has fewer settings set your grid up accordingly.

Upper tension

- The goal of the tension disks is to put an even amount of pressure or tension on the thread so it flows to the needle at a consistent rate. The dial (screen adjustments on computerized machines) to adjust your tension disks is usually numbered from 0-10 with 0 being the least amount of pressure being applied and 10 being the most amount of pressure applied. A balanced stitch is not the middle number on your dial. It is a stitch that is even top and bottom and the threads meet in the middle of the fabric layers. Bobbin thread will not be pulled to the top and top thread will not be pulled to the back with a balanced stitch. Stitches will be smooth, without puckering and the seam will be strong and secure even when stressed a little. The seam will feel smooth on both sides of the stitch.
- Machines are adjusted at the factory to have even tension at the middle number of the tension dial with the same thread in the top and the bobbin. This is often designated on the dial with a red dash. Under normal wear and tear of use the tension disks deteriorate and adjustments to the upper tension disks help to keep the stitch tension balanced. When using different weights of thread in the top and bobbin, adjustments to your upper tension are necessary to make a balanced tension stitch. The amount of adjustment necessary to the upper tension disk will vary from machine to machine. There is no magic number for a specific thread combination. What works on one machine may not work on another machine. When threading the machine or changing the tension adjustment the tension disks must be in the open position. This is achieved by raising the presser foot.

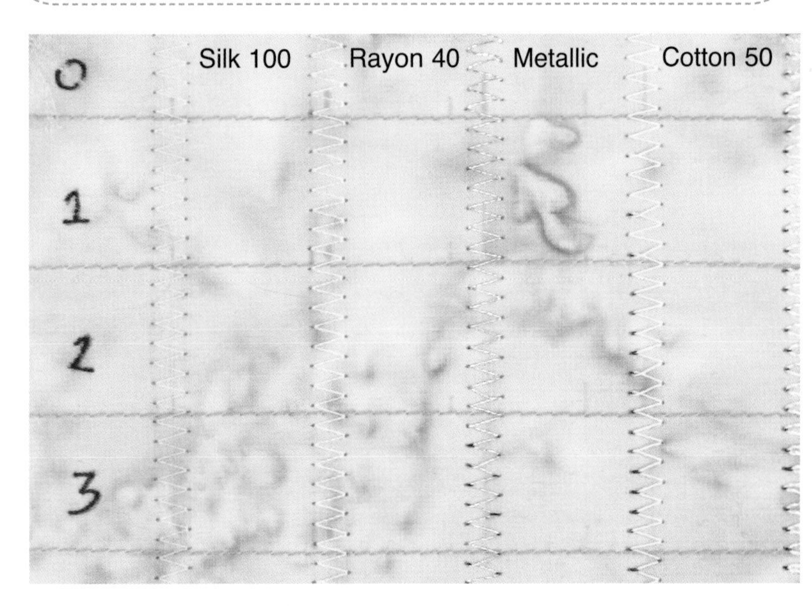

3 Select specialty threads and choose the needle down option if available. Begin at the top of the grid with the upper tension disks set to '0'. Stitch to the first grid line.

4 With the needle in the down position, lift the presser foot, adjust the tension dial to the next higher number. Lower the presser foot and stitch to the next grid line.

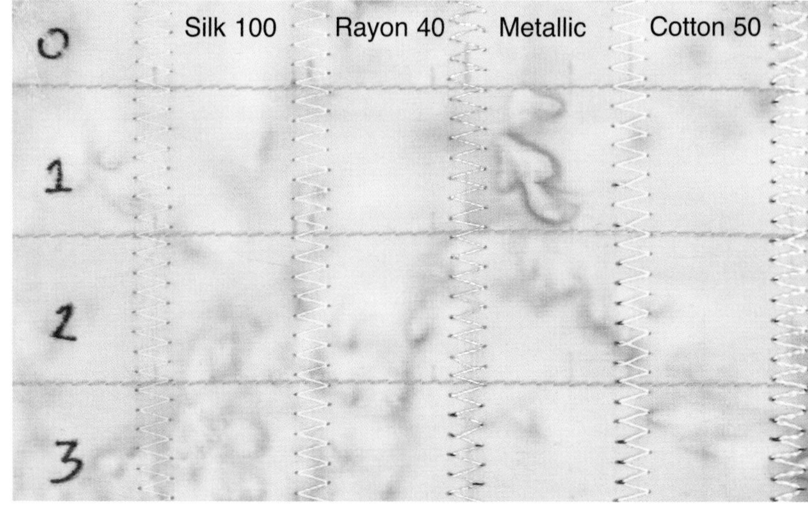

5 Repeat step 4 working through all the tension settings on your machine. Repeat the process using each of the specialty threads you would like to understand better. Write the name of the thread next to each sample.

take a closer look

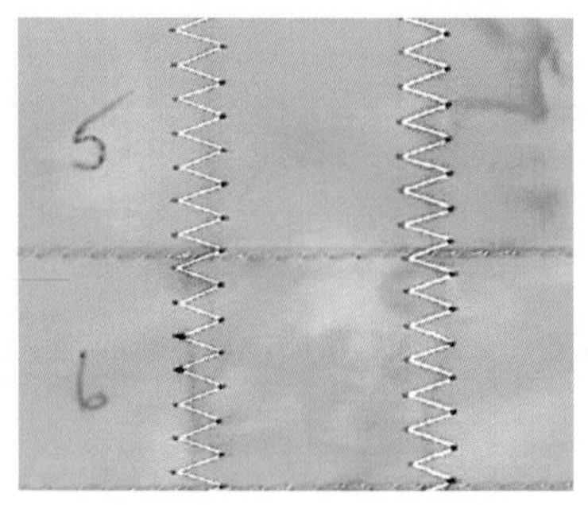

A closer look at our sample using a metallic thread reveals a 'normal' tension setting of 5 is too tight. The bobbin thread is visible.

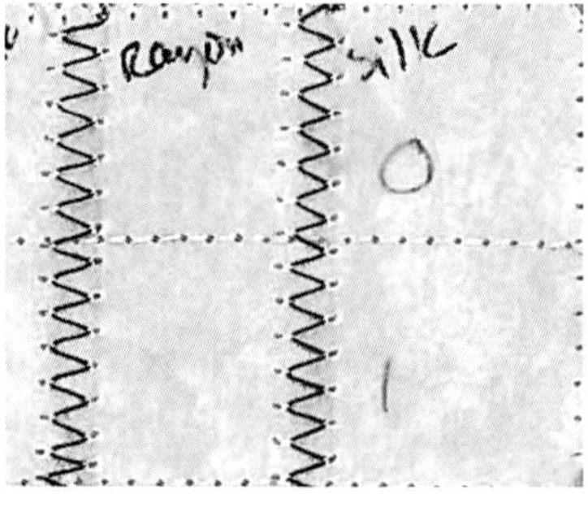

Look at the back of your sample. At a '0' tension setting it is generally expected to see the upper thread pulled to the back of the sample.

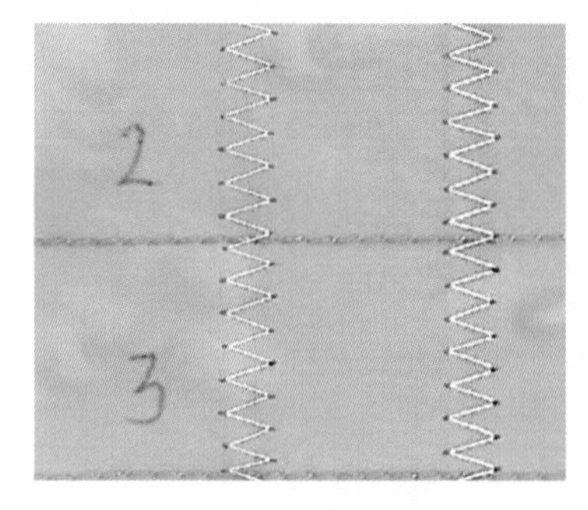

Our sample's balanced tension for the metallic thread is between 2 and 3 or about 2.5.

bobbin tension

Adjusting the bobbin tension often strikes fear into the heart of the most steadfast stitcher but it is truly a simple task and very rarely causes any problems. Adjusting the bobbin tension is not generally necessary for routine sewing tasks but is crucial to accommodate thick threads wound on the bobbin for the cable stitch and for achieving unique stitch effects such as a mock Cretan stitch, two color programmed decorative stitches, whip stitch, feather stitch, and moss stitch. Refer to Sewing with Ultra Heavy Threads on page 58 for using thick threads in your bobbin. Refer to Bobbin Case on page 16 to learn about adjusting the bobbin tension on a vertical and a horizontal drop-in bobbin case.

tip

- It is often recommended that you purchase a spare bobbin case to be used specifically for making bobbin adjustments. While this is good advice, bobbin cases are generally expensive and we do not think it is necessary. After you have practiced loosening and tightening your bobbin tension a few times you quickly learn what feels right and can easily adjust it back to a balanced position.

piecing

With these piecing samples you can see and feel the difference a thread's thickness makes in your seams. Ultra fine and fine threads combined with a sharp needle will improve your quilt piecing accuracy.

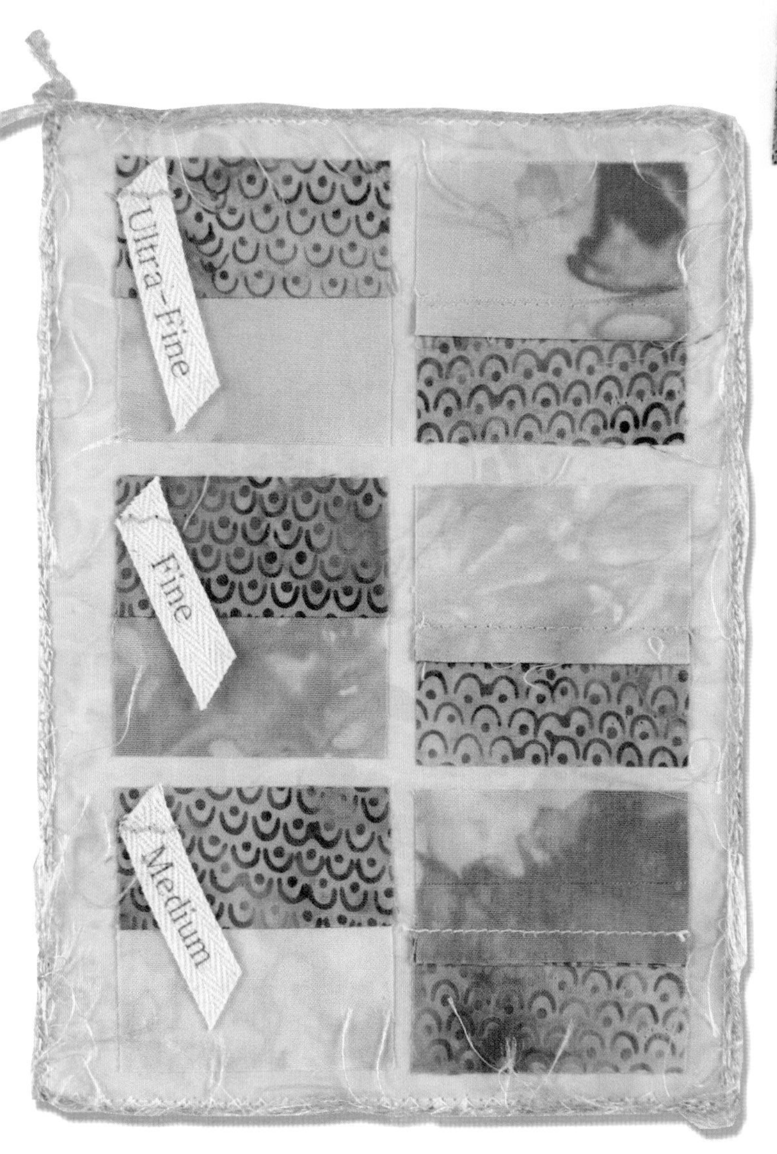

Edge finish: couching with a zigzag stitch

1 Cut a 1-1/4"-wide strip from two different fabrics. Cut each of the strips into 3, 4" long pieces. For the first sample set use the ultra fine thread to piece two of the strips together using a 1/4" seam. Press the seam to one side and cut the sample in half to make two samples each 2" wide.

2 Repeat step 1 using a fine thread and again using a medium weight thread.

3 Fuse the samples, one right side out and one seam side out, to a page base in order of their thread weight, lightest to heaviest as shown on the Passport page.

mock cretan stitch

& two color effects with programmed stitches

Adjusting the bobbin tension and using a simple zigzag stitch results in a stitch that looks very similar to a hand worked Cretan stitch. The Cretan stitch is made of open, looped stitches sewn right and left of a center 'line'. It may take a few tries to achieve the distinctive look of this stitch.

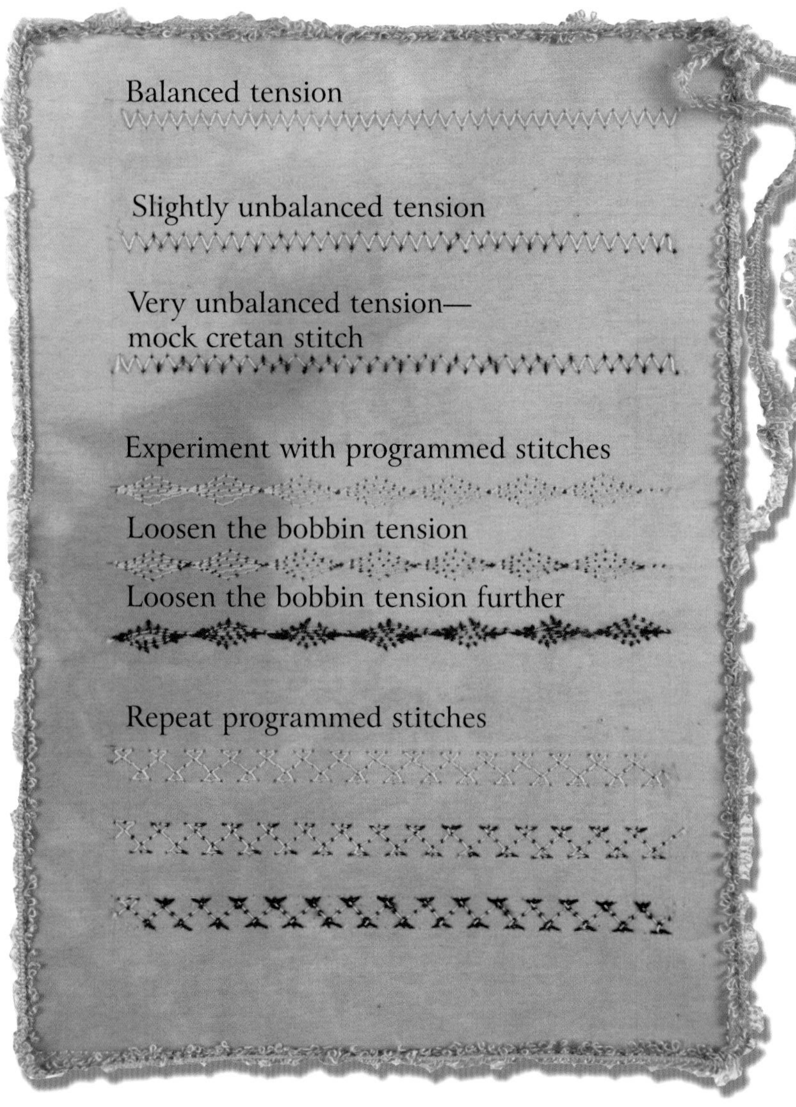

Edge finish: yarn couched with a zigzag stitch

Machine set up:
Normal

Tension:
Balanced or slightly loosen upper; loosen bobbin

Needle:
Embroidery 90/14 or sharp 80/12

Top Thread:
Any decorative thread

Bobbin:
Contrasting decorative thread

1. Stitch a row of zigzag stitches with a balanced tension.
2. Loosen the bobbin tension just enough to begin to bring the bobbin thread to the surface and stitch a second row of zigzag stitches below the first row. These stitches show a pop of bobbin color for an interesting effect.
3. Loosen the bobbin tension further to bring more of the bobbin thread to the surface and stitch a third row of zigzag stitches. This stitch resembles a Cretan stitch.
4. Experiment with some of your machine's programmed stitches. Stitch a row of the stitch with a balanced tension.
5. Loosen the tension slightly to bring up dots of bobbin color as you stitch another row.
6. Loosen the tension to bring up more of the bobbin thread for unique two color patterned stitches.
7. Repeat using a variety of your machine's programmed stitches to see the effects you can achieve.

whip & feather stitches

The whip and feather stitches vary by the amount of thread pulled up from the bobbin and by how fast you move the fabric under the needle. These stitches rely on both a loosened bobbin tension as well as a tightened top tension. They are worked with the feed dogs dropped as in free motion stitching. Exciting effects can be achieved by working in contrasting thread colors, blending thread colors, and multi-color or variegated threads.

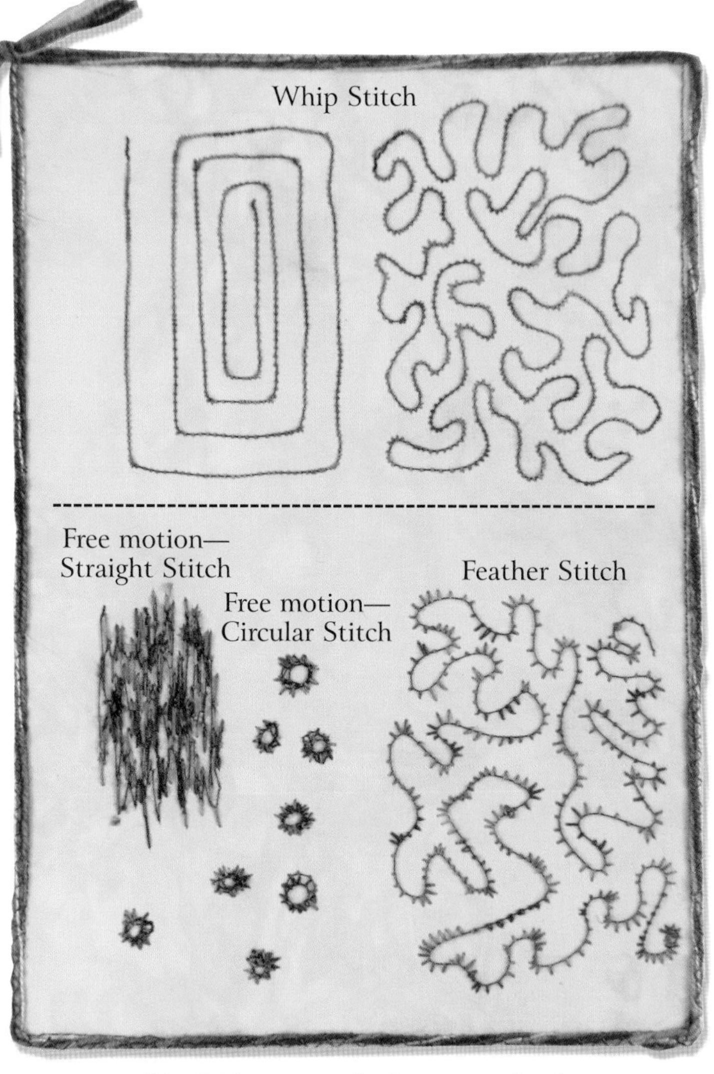

Edge finish: two strands of yarn twisted and couched with a zigzag stitch

Machine set up:
Free-motion
drop feed dogs

Tension:
Loosen bobbin and tighten upper

Needle:
Embroidery 90/14 or sharp 80/12

Top thread:
Any decorative thread

Bobbin:
Any decorative thread in a contrasting color to the top thread

whip stitch

The whip stitch resembles couching. The amount of thread pulled to the surface varies depending on how fast you move your fabric. For the most thread whip, run your machine motor fairly fast and your fabric very slowly. Experiment with the speed at which you move your fabric for different effects.

1. Layer your fabric with a stabilizer. Use an embroidery hoop if desired.

2. Stitch a variety of stitches and patterns. Experiment with the meander stitch, spiral, straight, and looping stitch.

feather stitch

The feather stitch works best with a fast motor and moving the fabric fairly fast with stitches that change direction quickly. Loosen the bobbin tension more than when creating the whip stitch and tighten the top tension. The goal is to bring the bobbin thread up to the top in big loops.

1. Layer your fabric with a stabilizer and an embroidery hoop if desired.

2. Loosen the bobbin tension even more than in the whip stitch. Try a variety of stitches and patterns such as the meander stitch, spiral, straight, and looping stitch.

moss stitch

The moss stitch takes the feather stitch one step further. The loops of thread brought up from the bobbin are exaggerated by moving the fabric in a circular direction or by changing direction frequently when using a straight stitch. After the stitching is finished the threads are secured by fusing a scrap of fabric to the back. The top thread is removed with a seam ripper or scissors leaving loops of thread on the surface.

Edge finish: cord couched with a zigzag stitch

Machine set up:
Free-motion
drop feed dogs

Tension:
Loosen bobbin and tighten upper

Needle:
Embroidery 90/14

Top thread:
Water soluble or fine polyester thread in contrasting color to bobbin

Bobbin:
Any decorative thread.

1 Cut your fabric large enough to use in a machine embroidery hoop. Cut a piece of heavy tear away stabilizer the same size.

2 Draw a heart shape on your fabric and layer the fabric and stabilizer together. Insert in the hoop.

3 Begin stitching a circular feather stitch just inside the heart pattern.

4 Continue to stitch the inside of the heart in the same circular pattern until you are satisfied with the density.

5 Take the fabric out of the hoop and turn it over. Fuse a piece of muslin to the back of the stitching with a fusible web

6 Remove the top thread by picking it out with a seam ripper or dissolving in water if using a water soluble thread.

tips

- The bobbin thread is pulled up more readily when the stitching changes directions or curves.
- If the upper thread begins to break, loosen the top tension a bit and loosen the bobbin tension a little more.
- Working with a heavier stabilizer and/or an embroidery hoop will help to control fabric distortion when dense stitching is desired.

edge finishes

This set of Passport pages is great for quick inspiration when you are finishing a project. Or, you can just as easily try your own edge finishes on your finished Passport pages.

Edge finish: ribbon stitched with a straight stitch

Edge finish: yarn couched with a zigzag stitch

getting started

1 Cut fabric into twelve 2-1/2" x 6" pieces and four 3" x 6-1/2" pieces.

2 Cut batting into six 2-1/2" x 6" pieces and two 3" x 6-1/2" pieces.

3 Sandwich each of the 2-1/2" x 6" batting pieces between two 2-1/2" x 6" fabric pieces. Pin layers together to create mini quilts.

4 Choose a different edge finish technique to try on each mini quilt.

5 Arrange your mini quilts on a fabric page base and layer of batting. Stitch down on one side using a straight or zigzag stitch.

Corners

Some of these techniques cause the corners of your piece to become slightly rounded. If you prefer square corners thread each corner with a thread tail. You can hold on to these tails as you turn the corner to prevent the fabric corner from being pulled down into the throat of your stitch plate. Remove the thread when finished.

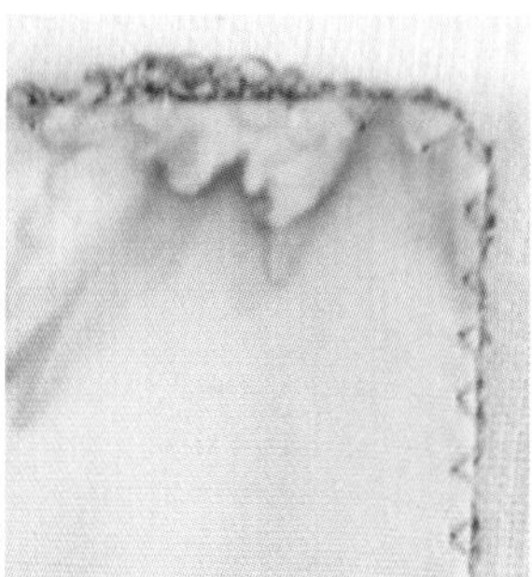

The simple edge created with a pillowcased or faced edge gets additional texture with yarn couched at the edge. Use a blending or contrasting thread for different looks. Sew the 3" x 6-1/2" fabric pieces, right sides together, around three sides and turn inside out. Insert batting and press before couching the yarn. Refer to Couching on page 90.

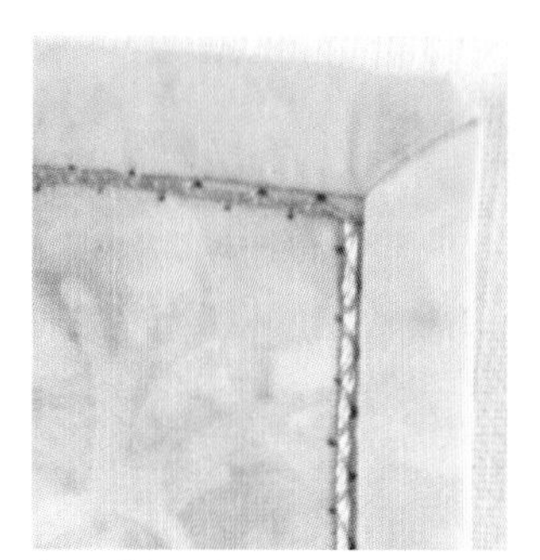

A traditional quilt binding is sewn on the front of the mini quilt, pressed over the edge to the back, and hand stitched. We added a couched rayon heavy thread as an accent. Multiple rows of fibers could be added for more texture or color. Refer to Couching on page 90.

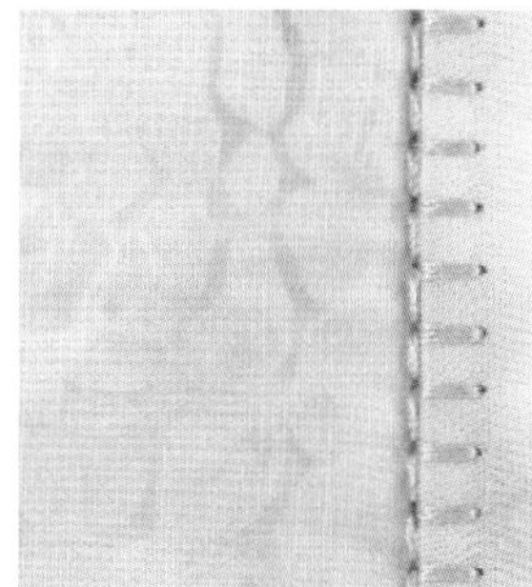

For a different effect with a traditional binding, sew the binding from the back of the mini quilt. Press forward over the front of the quilt and use a decorative machine stitch to secure it. It is helpful to secure the binding with 1/4" fusible web tape before stitching.

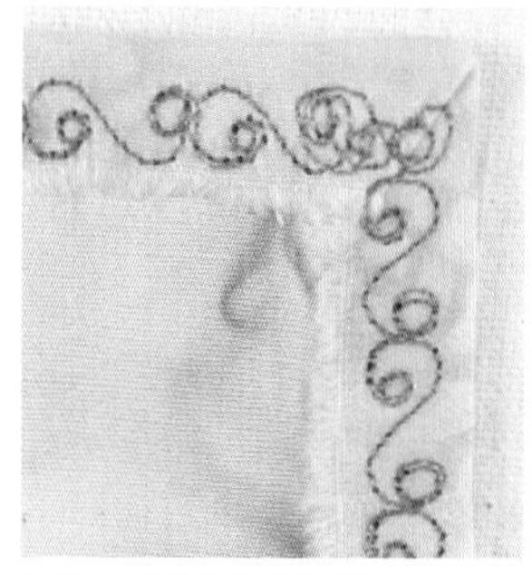

A ripped fabric binding is a casual funky edge finish. We used a 1" piece of fabric and ripped it instead of cutting it. Fold the fabric in half over the edge of the mini quilt and stitch with a zigzag, straight, or decorative stitch. Ribbons are also fun to use.

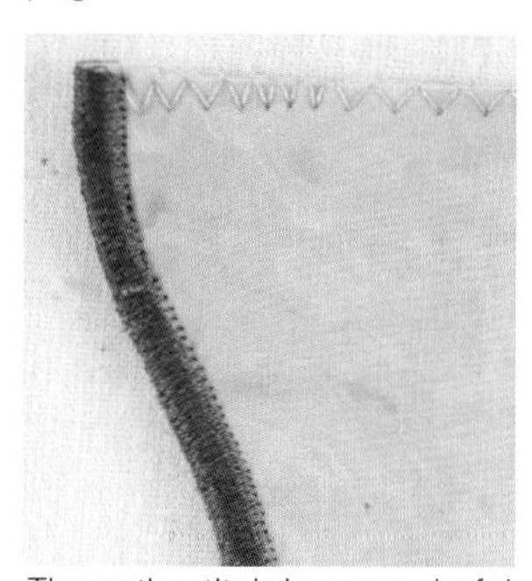

The satin stitch is a wonderful edge treatment to add to straight or curved edges. It requires some additional steps as outlined on page 38.

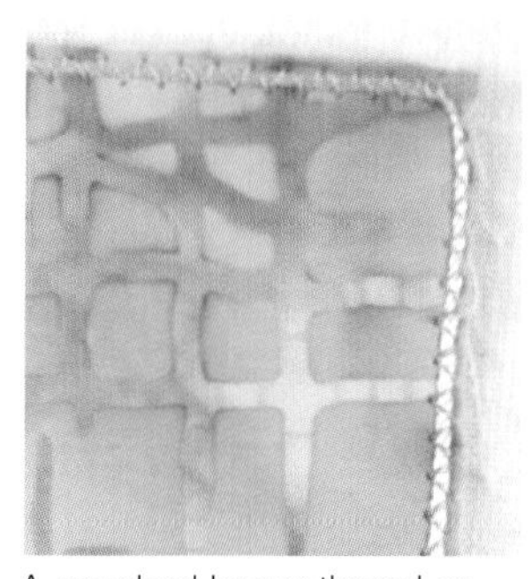

A couched heavy thread or cord gives a simple but fun edge finish that frames your work. Try couching down two heavy threads in the same or different colors. Refer to Couching on page 90.

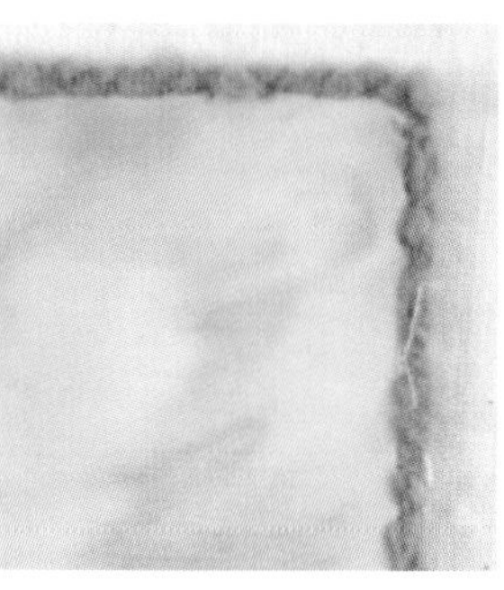

A heavy thread or chunky yarn couched on the edge can be just the ticket to highlight a small piece. Refer to Couching on page 90.

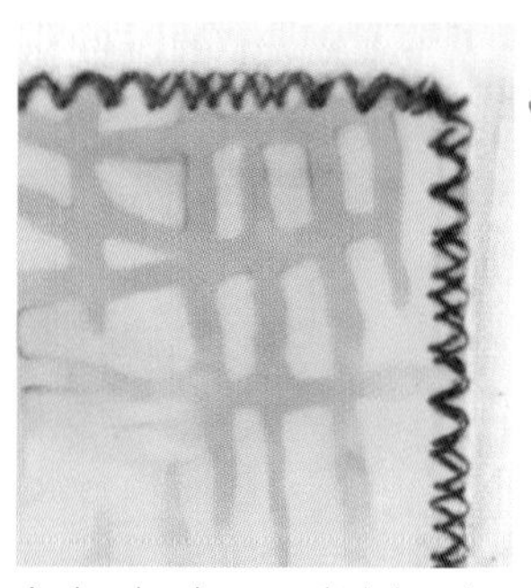

A simple zigzag stitch in a heavy rayon thread around the border gives a fun and simple finish. The zigzag stitch is set at a medium width and length. Decrease your stitch length and stitch around the edge again for extra coverage if desired. With a medium weight thread, stitch around the edge three or more times to reach the desired coverage.

edge finishes

We have given directions for a curved edge finish. Use the same steps for a straight edge finish.

1 Draw a curved line on a 3" x 6-1/2" fabric piece. Use the template below or draw your own.

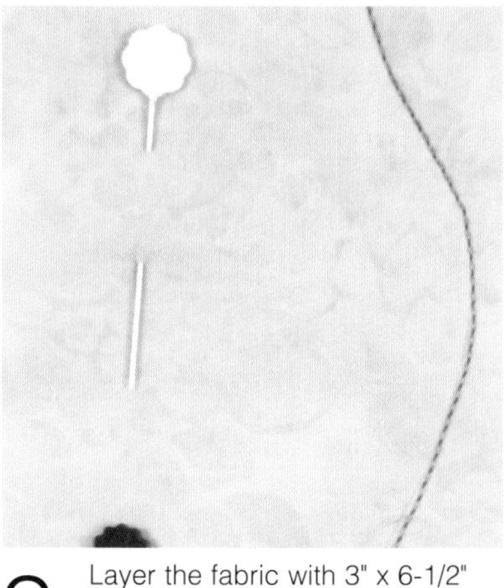

2 Layer the fabric with 3" x 6-1/2" batting and backing fabric pieces. Stitch on the drawn line with a straight stitch. This will help stabilize the fabric.

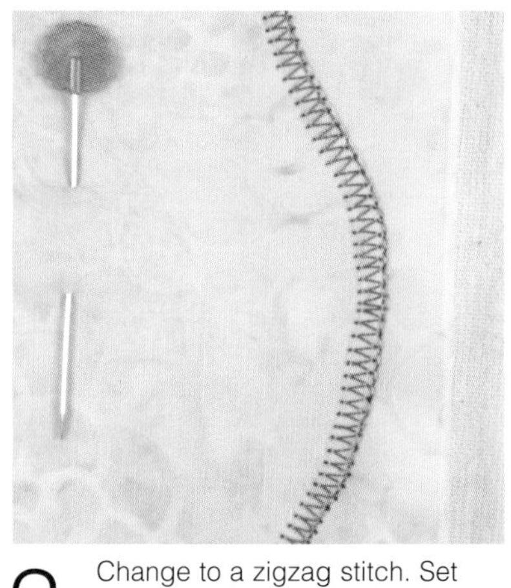

3 Change to a zigzag stitch. Set the stitch length to slightly longer than a satin stitch (between 1 and the satin stitch setting) and the width to just under the final zigzag width you want. Stitch along the edge of the previously stitched line.

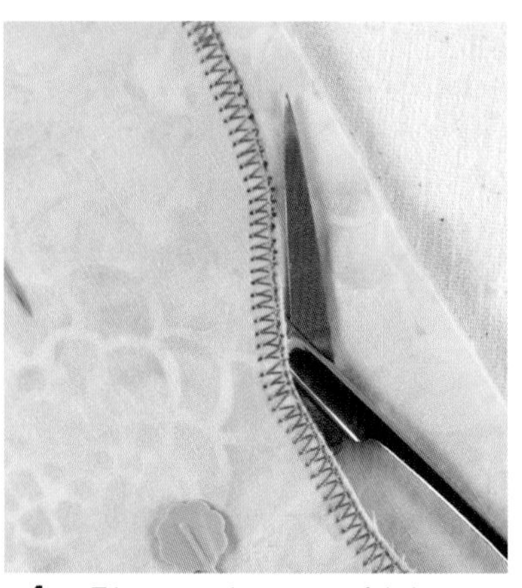

4 Trim away the excess fabric as close as possible to the zigzag stitching. Don't worry if you accidently snip a few threads.

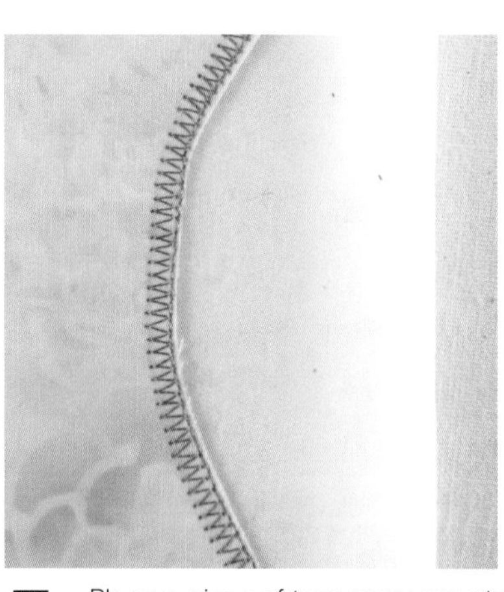

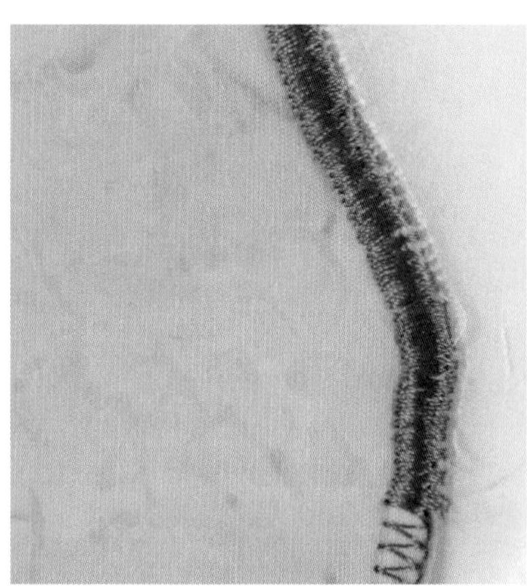

5 Place a piece of tear away or water-soluble stabilizer behind the sewn piece. Some of the stabilizer should show on either side of the edge. Set your zigzag to the satin stitch length and as wide as possible or desired. Stitch over the previous stitching. Position the needle so the right swing of the needle is just off the edge of the fabric. Remove stabilizer and press.

Curved Edge Template

Thread Sampler

Create your own sampler of stitches using a variety of threads to audition their effects. This sampler uses free motion stitching and a variety of techniques including 2 threads in 1 needle, twin-needle stitching, and zigzag free motion.

the techniques

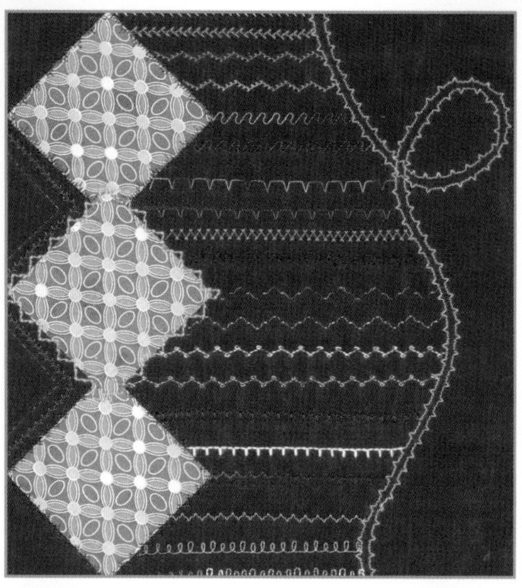

quilting out of the ditch

appliqué

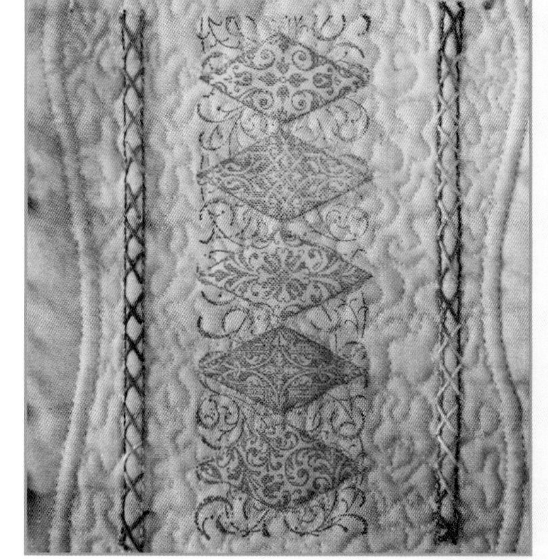

trapunto

sewing with
ultra heavy threads

hand look quilt stitch

Thread in Quilting

Quilting is all about the thread. Without thread we can't hold our gorgeous fabric bits together. In this stage of our journey we are going to look at the role of thread in our quilting. Quilting has changed in the last decade as quilters explore contemporary and art expressions in addition to honoring the traditional roots. Our thread choices are no longer limited to cotton in demure colors. Heavy threads make a colorful bold statement in quilting and appliqué. Conversely, threads can be almost invisible and let the fabric textures be the center of attention. We hear a lot about the rules of quilting like perfectly balanced stitching, perfectly even stitches, and no crossing of meander stitch lines. Rules are great teaching tools. They are a code that helps us to learn, but this is your quilt. You make the rules.

quilting out of the ditch

On the following pages you will find an assortment of ideas to jump-start your quilt journey. Quilting, or the act of stitching three layers together, is often done either in the ditch or with free motion. These portfolio pages offer a wide range of options between those two extremes. They are filled with as many threads and ideas we could squeeze into a square. The emphasis of this exploration is on experimenting and challenging yourself to broaden your use of thread.

Refer to page 21 to prepare your quilt for quilting.

decorative stitch sampler

Stitch a sampler of designs to see what works best for quilting. Check to see if you can live with the way the stitches look from the back. Not all machines will be rich in stitch variety, but that shouldn't stop you from creating alternative quilt stitches. The Blind Hem stitch is available in some variation on most machines and can become a great wild vine when used in parallel rows in opposite directions. Our green, thorny vine was inspired by the quilts of Jane Sassaman. It is particularly bold stitched out in heavy weight wool thread. The decorative stitches act to lightly appliqué the raw edges of the three squares. Their burden of securing the squares in place is lessened by first stitching the perimeter of the squares with an anchoring course of fine thread in a narrow zigzag.

around the block, mimic or askew

Mimic: The top two mimic blocks repeat the basic lines of the pieced block. The top right example is the pieced block where ditch quilting is emphasized with a heavy thread stitched in a pattern instead of a straight line. It is also done as a continuous line, without having to stop and trim threads repeatedly. The top left setting square references the quilted concentric squares. The same heavy thread is used with a decorative stitch in a continuous line, surrounding a small central square stitched in red.

Askew: The same design is used in the bottom two blocks, but is turned on point. Individual squares are stitched in the adjacent setting square with the outer square framed in a decorative stitch. On larger quilts, stitching small, independent elements can be cumbersome since you will have to manipulate a large quilt in the machine space.

tips

- The walking foot is a good choice for parallel lines as it often has a quilt bar that can be connected for guiding the parallel rows of stitching.
- Check for colorfastness. Stitch on fabric scraps with your desired threads. Soak the sample in water and check for dye transfer.

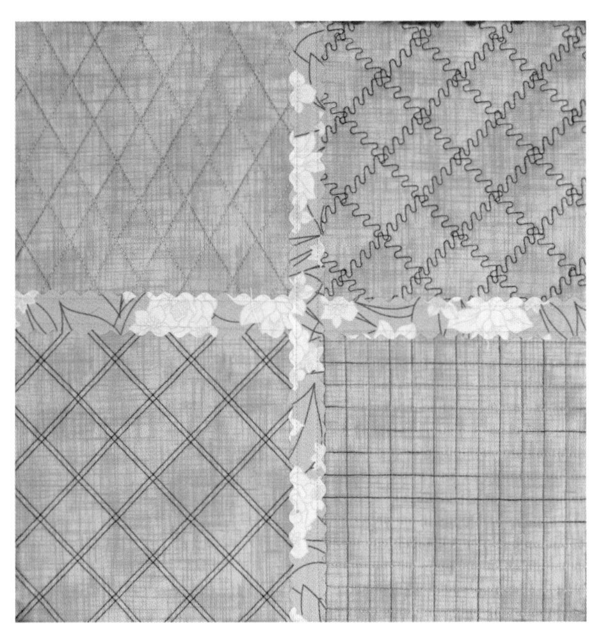

grids

Hanging Diamonds: A set of lines was marked using the 30-degree mark on the ruler. These lines were then spaced the width of strips of masking tape. The lines were stitched in a heavy weight variegated thread.

Faux Houndstooth: The look of houndstooth check can be achieved with a wavy decorative stitch in crossed gridlines.

Twin Needle Quilting: A twin needle works with two spools of thread on the top held in place by one bobbin thread. The quilt back appears to have a zigzag stitch from the back with parallel lines on the front. If you don't mind this on a quilt back, or you are making a wallhanging, this might prove to be just the quilt finish you need. The block was marked with single chalk lines and was stitched with one of the twin needles to either side of the line.

Variegated threads: Multi-color threads can add a more dynamic feel to a straight line.

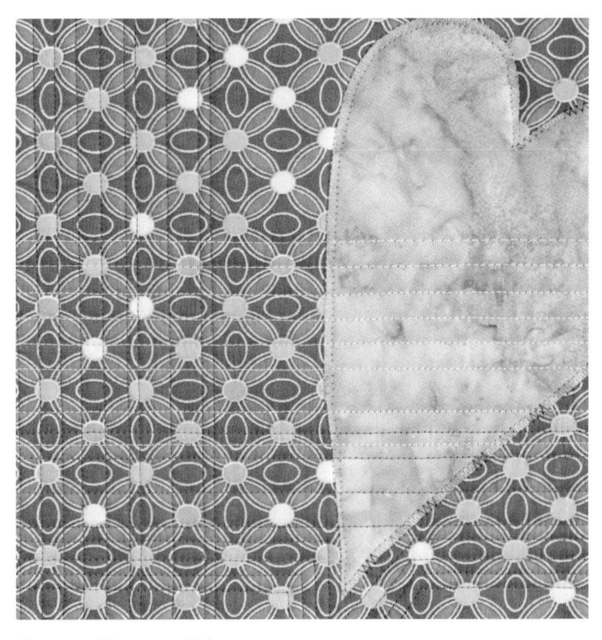

channeling quilting

Blending lines by weight: The lengthwise lines are stitched to blend by using an ultra fine thread in a blending blue color. The result is the suggestion of a line but the stitches are not well defined.

Blending lines by color: The lines moving across the heart demonstrate a few things. The top lines are done in a medium weight thread of a blending color. These stitches do not disappear, but they do not call out for much attention. The top line stitches also have a longer stitch length than the lower lines. Longer stitches are more clearly defined.

Accent line: Choosing a thread in a lesser used fabric color can add a 'pop' moving across the quilt. In this case, a neon green was chosen.

tip

low tack taped lines

- This is a great practice opportunity. Try a few different widths of masking or painter's tape. Lay out a small pattern and stitch around it. You will get a feel for the lines that remain after stitching around the tape. When turning corners, try to take an extra stitch in the corner so it does not get rounded off. Take locking stitches at the beginning and end of lines, as there can be lots of small runs of stitching. Tape is a great way to achieve straight lines on a large quilt. Do several measurements along a line to be sure lines are equidistant and true.

geometric pivots

Echo Pivots: When 'echoing' angular edges of quilt motifs (as with the pointed fabric appliqué shown), use the width of your presser foot as a guide. Alter the spacing by adjusting the needle position left or right. Appliqués should be fused and/or anchor stitched as the pivot stitching is not designed to catch all sides of the appliqué securely.

Triangle Pivots: Before trying this technique on a full size quilt, try it on a good sized fabric sample or a small project to get a feel for the movement you like best. To pivot the quilt at the point of a triangle, leave the needle in the quilt when you get to a corner. Lift the presser foot and turn. Lower the presser foot and continue stitching until the next pivot. With practice, you can avoid completely turning the quilt by making use of your reverse button. You can usually keep the quilt moving forward in one direction. At pivot points, make a slight pivot and then REVERSE to the next pivot point, where you re-engage the presser foot and move forward again.

couching

Seam lines and appliqué edges are great opportunities for a couched line of stitching. The couching stitches can be nearly invisible with monofilament thread in a zigzag or they can be exaggerated with a heavier thread. Take care to use a strong thread as it is doing double duty in couching and quilting. Refer to Couching on page 90.

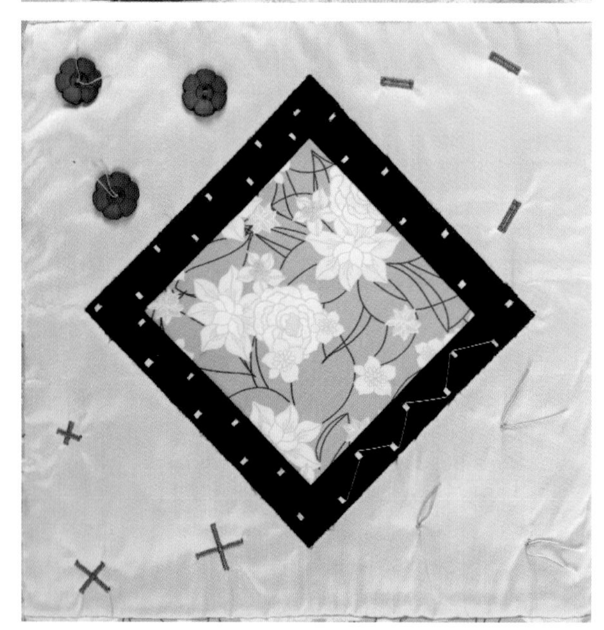

tacking & tying

Button Tacking: Shankless buttons are stitched in place with a strong thread.

Satin Stitched Bars: Use matching top and bobbin threads to stitch either a staggered row of short bars, as in the dark blue, or scattered crossed bars or satin stitches.

Tying: Thread a large eye straight or curved needle with embroidery floss, perle cotton, or a similar heavy weight thread. Push the needle through to the opposite side of the quilt and back to the starting side within an 1/8", leaving 4" tails at either end. Tie the thread ends securely.

Buttonhole Tacking: Considered a four-sided bar tack, this method is most effective with basic buttonhole stitching and matching medium to heavy weight top and bobbin threads.

continuous line quilting

Inner Square, Hearts: The design we chose fits neatly in a 6" block. Trace your design onto Golden Threads™ paper. Stitch directly through the tissue pinned over the quilt block. Start stitching from any point along a line. This design can be stitched in fine to heavy thread. For alternate methods of transferring designs, refer to Marking Tools on page 16 for additional products. Follow manufacturer's instructions.

Outer Square, Loop Border Stencil: The stencil design is transferred to the fabric with a chalk pouncer. (Stencil source-Hariett Hargrave, HH8, www.quiltingcreations.com).

Side 1 (top)- Medium thread, free motion.
Side 2 (right) - Medium thread, sculptured or hand look quilting stitch.
Side 3 (bottom)- Heavy cotton thread, free motion.
Side 4 (left) – Wool thread, free motion

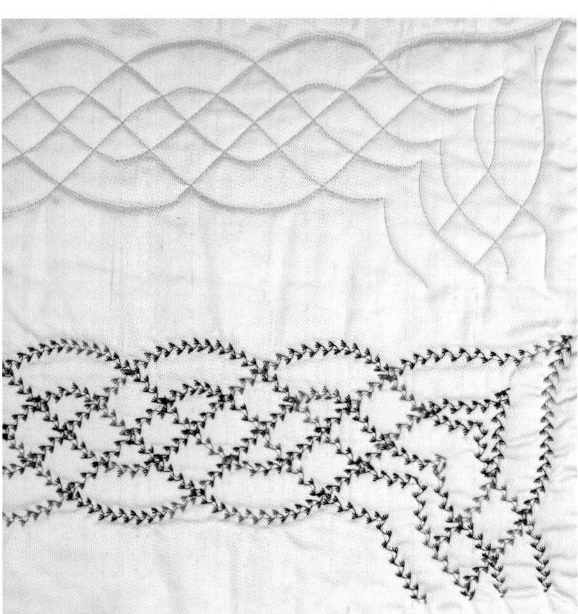

continuous line, decorative stitching

Transfer the design to the quilt top with your method of choice. Stitch one continuous line at a time with a straight stitch in an ultra fine thread. Re-trace stitched lines with a variegated rayon thread in a decorative stitch.

continuous line, bobbin stitched quilting

Transfer the design to the quilt top with your method of choice. Stitch one continuous line at a time with a straight stitch in an ultra fine thread. Turn the quilt over. With ultra heavy thread in the bobbin and fine thread on top, follow the anchor stitch lines to create the stencil pattern 'blindly' from the back of the quilt. This sample was worked with bobbin sewing a straight stitch, but heavier threads or alternate stitches could be used. Refer to Sewing with Ultra Heavy Threads on page 58.

tip

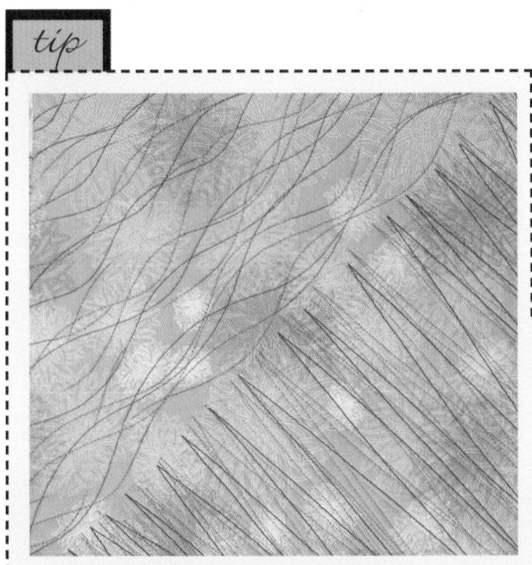

wavy & jagged line quilting

- Capture the look of free motion quilting with a walking foot. Stitch line movement can be anything from a calm flow to frantic energy. Feel free to cross lines and change up the threads.

free motion stitching

Free motion stitching is one of the best tools in the stitcher's toolbox of techniques. As its name implies, it offers an unparalleled freedom of sewing. If you can think it, you can most likely free motion stitch it. Quilting, thread painting, free appliqué, and thread writing are all possible with this versatile stitching.

free motion stitches

Make a stack of quilt sandwiches using scraps of batting and muslin. Use the same weight thread in the top and bobbin. Practice these exercises routinely and you will soon be a confident free motion stitcher. Some patterns may seem easier than others and once you feel confident with these patterns try vines, leaves, and swirly spirals. The sky is the limit.

Try This

- Backward and forward stitches squared off at the ends
- Backward and forward in a long zigzag
- Backward and forward using small curving lines
- Cursive e's and l's
- Loops and circles
- Simple shapes, squares, triangles, hearts, stars, flowers, vines
- Your signature
- Meander large then tighten to tight stipple (stipple is any stitch pattern that is stitched closely together)

emotive free motion stitching

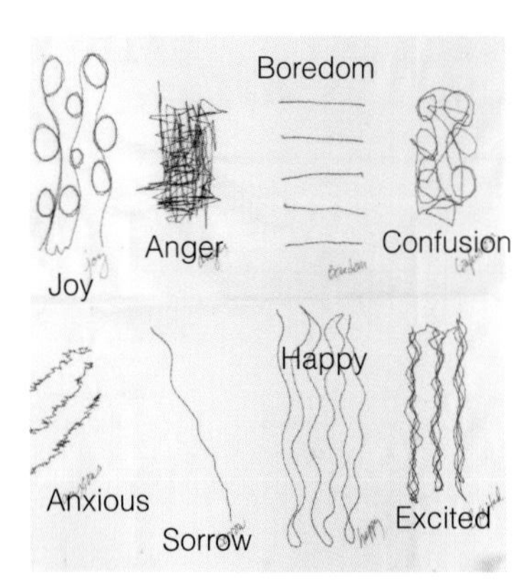

Lines can express emotion. This is true for both the drawn or stitched line. Your quilting line can affect the mood and energy of your quilt.

- Draw a grid on a 12" x 12" piece of muslin. Back the muslin with a stabilizer or make a quilt sandwich. In each section, stitch lines in a manner to depict a variety of emotions as seen in our sample. Do not over think this exercise.

Preparing yourself and your machine

- To set your machine up for free motion stitching drop or cover the feed dogs and insert a darning or free motion foot. When free motion stitching you are taking over the job of the feed dogs and become part of the tension system. Moving your fabric inconsistently or with jerks can cause skipped stitches, unwanted bobbin thread showing, or breaking thread. Try loosening the top tension slightly to compensate for your style of moving the fabric under the needle. Refer to your sewing machine guide for directions in setting your machine up for free motion stitching.
- After your machine is set up, get yourself ready. First and foremost you need to be relaxed. A great way to practice stitch designs is to doodle. This will help you develop your own unique marks and rhythms.
- There are many quilters who share their 'secrets' for free motion success but the real 'secret' is to discover a pace and rhythm that works for you. There is no wrong way to stitch. The universal secret to successful free motion quilting is practice, practice, practice. The goal is not to achieve perfection based on someone else's definition of it but to achieve a look you like that is natural and comfortable to do. You will discover some stitch styles and patterns suit you better than others.
- One of the most common stitches is the meander stitch. In reality this is not the easiest stitch to master. Then someone throws in the "you can't cross your stitch lines rule". There is nothing more than the fear of not doing it right to stop us in our tracks before we sew one stitch. We are here to tell you we have crossed our stitch lines and the fabric of the universe did not rip apart. Quilting is supposed to be fun and if it works for you don't worry about pleasing other people.

specialty threads in free motion stitching

Edge finish: yarn couched with a zigzag stitch

1 Use an iron to press the fabric page in half lengthwise and then in thirds across the width to divide the page into six sections. Mark the creases with pins as a guide if desired.

2 Using six different types of threads, stitch a meander or other free motion quilting stitch in each section. We used an ultra light weight, a metallic, a medium weight cotton, a heavy weight cotton, a medium weight rayon, and a heavy weight rayon.

free motion stitch sampler

1 Prepare a fabric page with batting and tear away stabilizer.

2 Using a cotton or rayon thread, try different free motion stitch patterns like circles, loops, squares, vines, or hearts...whatever you love.

Edge finish: yarn stitched with a straight stitch

Play with your stitches

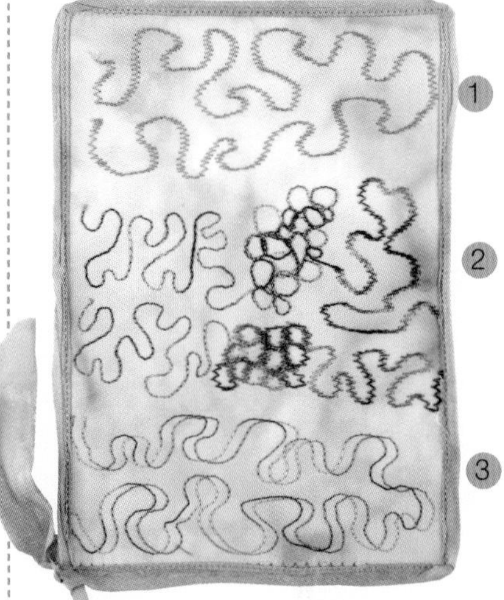

Edge finish: ripped strip of silk stitched with a straight stitch

1. Layer a fabric page base with batting and tear away stabilizer. Use pins or draw a fine line to divide the page into three sections. Try free motion stitching with a zigzag stitch. Check to make sure you use a zigzag width narrow enough so the needle can easily clear the sides of your free motion foot.
2. Chose a size 90 embroidery needle. Thread two threads through the needle at the same time. Check with your sewing machine guide for threading directions. Try two different colored threads or use the same variegated thread through both needles for a mottled effect. The top sample shows the same variegated thread through both needles. The bottom sample shows two variegated threads, one in shades of pink and the other in shades of pink and orange.
3. Choose a twin needle and free motion stitch to get a shadow or ribbon effect while free motion stitching. With a wide free motion foot you can try a narrow zigzag stitch. *Note: There is no pivot point with two needles so you cannot rotate the fabric while the needle is in the down position.*

appliqué

Appliqué is an incredibly versatile technique and lends itself to traditional, contemporary, folk, or funky styles. If you are seeking to develop your own style and rely less on traditional patterns, appliqué is an easy way to begin branching out into art quilting.

Edge finish: yarn couched with a zigzag stitch

Edge finish: four strands of silk yarn couched with decorative stitch

getting started

Machine appliqué can take on many forms. Use fine threads and a hem stitch for a nearly invisible appliqué or use a satin, blanket, or other decorative stitch to create a thread focused edge.

One of the most popular techniques today is raw edge appliqué. In this technique the fabric motif is fused to the base fabric and stitching is done for both security and embellishment. We have used a simple heart shape because it is a wonderful appliqué shape for beginners. Use the template below or design your own.

1 Layer a fabric page base with a stabilizer. Choose a fabric large enough to accommodate the number of shapes you want to appliqué. A 4" x 11" piece is large enough for six hearts if using the pattern provided. Trace the heart pattern on a piece of freezer paper and cut out. Iron the freezer paper pattern to your appliqué fabric. Trace around the pattern with a pencil. Remove the freezer paper and repeat to mark the number of shapes needed.

Satin Stitch | Blanket Stitch | Free Motion Stitch

Invisible Hem Stitch | Granite Stitch | Triple Stitch

2 Iron a piece of fusible web on the back of the fabric and cut out the appliqué heart shapes. Arrange multiple hearts, fusible side down, on your fabric page base and fuse in place with a hot iron. Try different stitches to secure the fabric edges. Refer to the Stitch Dictionary on page 22.

- Apply the fusible web to your fabric before you cut out your appliqué shapes. This will help prevent the raw edges from raveling.
- When using the invisible hem stitch decrease the width and length of the stitch and use one of the ultra fine threads for an almost invisible stitch.

appliqué

The free motion hatching stitch is made up of thread lines laid down closely for a heavily embellished edge. It is a great place to take advantage of the wonderful properties of multi-color threads.

Edge finish: yarn couched with a zigzag stitch

- When working with multi-color thread it is best to work around the entire piece lightly at first and then go over the first stitching one, two or even three more times to achieve the desired density. This blends the colors and eliminates the stripes of color due to the color changes in the thread.
- Hand Appliqué: Try using one of the ultra light weight or light weight threads the next time you appliqué. These fine threads slide through the fabric smoothly and it is almost impossible to see the fine stitches.

1 Layer your base fabric with a heavy stabilizer or use two layers of stabilizer to discourage distortion. Apply fusible web to the back of your appliqué fabric. Draw a heart motif or use our pattern on page 49. Cut out and fuse the shape to your base fabric.

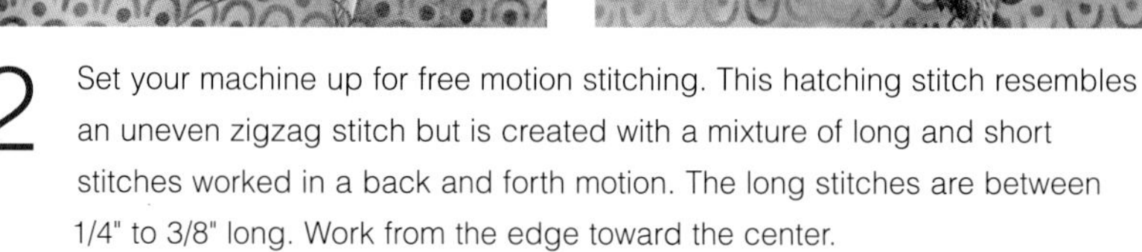

2 Set your machine up for free motion stitching. This hatching stitch resembles an uneven zigzag stitch but is created with a mixture of long and short stitches worked in a back and forth motion. The long stitches are between 1/4" to 3/8" long. Work from the edge toward the center.

Spring Quilt

This cheerful spring quilt showcases fun graphic prints in a quick and easy pattern. Twin needle quilting in a diagonal trellis pattern support the spring garden feel.

Twin Needle Stitching

trapunto

Subtlety is the key to successful trapunto. When the motif outline stitches disappear into the background, the focus is on the sculptural effect and not the stitched line. In this case, the thread takes on a supporting or special effects role rather than the lead.

Edge finish: couching with yarn and pearl cotton

Edge finish: rickrack sewn on with a straight stitch

getting started

Machine set up:
Free motion
drop feed dogs

Tension:
Balanced

Needle:
Sharp 10
or the smallest needle that works
with fabric and thread choice

Top thread:
Ultra fine, monofilament or
water soluble

Bobbin:
Fine

Fabric Size:
Trapunto tends to shrink a page significantly. Cut your fabric at least a 1/2" bigger in length and width or mark your page size on a larger piece of fabric. Complete the stitching and then trim it to size.

Batting choice:
A higher loft batting, often used for the first layer, fills a trapunto motif and lets it stand higher. Use what you like or have on hand.

tip

- Practice drawing the design on paper before stitching. You can also draw the motif onto your fabric with a removable marker.

1 Layer fabric over batting. Use a few pins to secure. (Spray basting is not recommended, as you will be trimming away a significant portion of this first batting layer.) Free motion stitch one or more hearts on the fabric. This step might be thought of as basting. It marks the design and holds the initial batting in place to be trimmed.

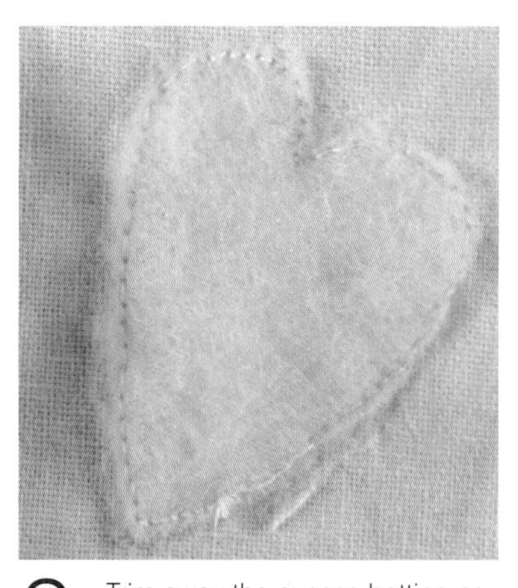

2 Trim away the excess batting as close as you can to the stitch line. Be careful not to cut the stitches or your motif outline will disappear from the front of the fabric. Use blunt tipped scissors or duck billed appliqué scissors to help avoid cutting the fabric. If you do cut stitches, you can alter or re-stitch a portion of your design.

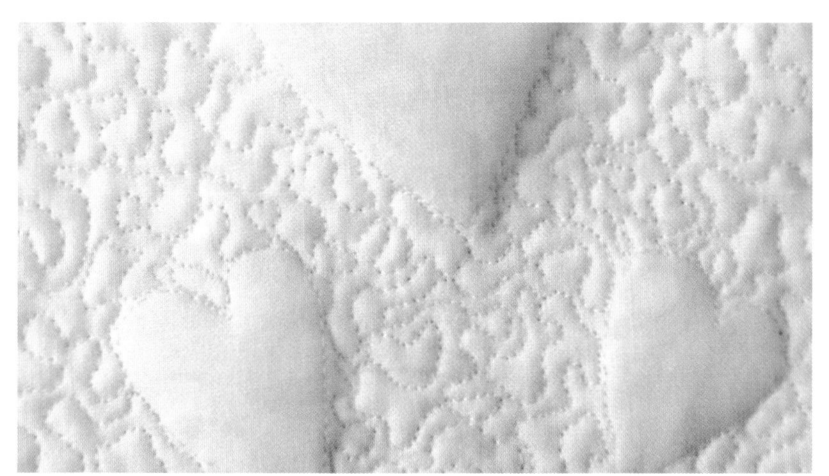

3 Add a second layer of batting behind the trimmed layer followed by either stabilizer or fabric. Fill the background and compress the second layer of batting by moderately stitching around the hearts. An ultra fine thread in a blending green color is used to stipple stitch the background and make the hearts stand out.

Edge finish: three strands of thread zigzag stitched

Stitching Options

Free motion stitching is not your only background quilt stitch option. You can stitch with the feed dogs up in a grid or other pattern. Stitching should be fairly dense in order to 'pop' the trapunto design. Using a background quilting stitch too close to the scale of the motif itself lessens the contrast and detracts from the effect.

When stitching the fill stitch, you might also 'dip into' the motif for some accent stitching and continue the accents into the fill stitches.

It's worth experimenting to find the look you want. Monofilament thread gives an equally subtle effect. You can barely see the stitching, whereas with a blending thread, you tend to see the stitching upon looking closer.

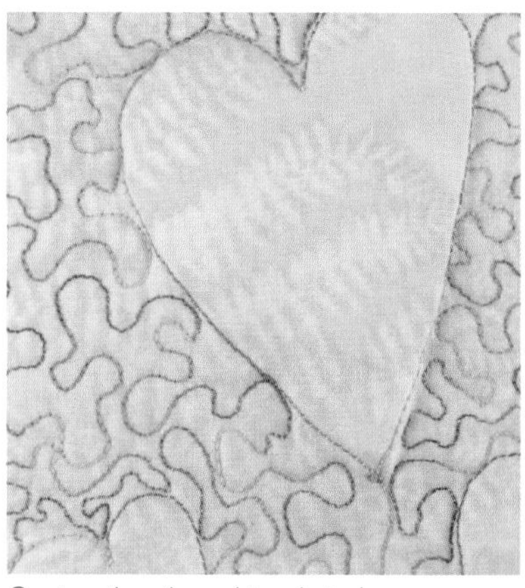

Contrasting thread tends to lessen the focus on the 3-D effect of the trapunto. It also makes the motif outline more obvious.

Edge finish: yarn couched with heavy thread

Fabric pattern trapunto

To give a fabric design a trapunto effect, choose a closed shape to stitch around and follow the steps listed for trapunto.

trappliqué

Edge finish: repeated zigzag stitching

Trapunto appliqué, referred to as trappliqué, is particularly effective as the water-soluble thread is stitched just inside the motif, followed by your appliqué stitch of choice just outside the appliqué. The water-soluble thread is then rinsed away.

1 Stitch the fabric appliqué to a layer of batting and the background fabric using water-soluble thread on top and fine white thread in the bobbin.

2 Add appliqué stitches around the motif and the batting, overlapping the water-soluble stitching.

3 Add fill stitches around the motif. Rinse away the original water-soluble stitching by slightly rubbing the stitching with your fingers under tepid water. Air-dry or use the dryer. DO NOT press dry with the iron or you will compress the trapunto. Trim the edges to size, add backing, and finish the edges

These steps were done using 2-1/2" x 3-1/2" fabric pieces. They were lightly stitched in place for demonstration. You could stitch a single finished piece to a background page.

Trappliqué notes

Green wool thread used for the stem was pre-stitched on the fabric before any batting was added so as not to trap the first batting layer. Pre-planning your stitching when there are multiple steps is important.

Lots of stitching caused our page to shrink up. The shrunken page was trimmed and stitched to the top of a new, properly sized background, creating a framed effect.

Edge finish: repeated zigzag stitching

faux corded & channel corded trapunto

Try these three ideas for achieving corded trapunto effects with a twin needle.

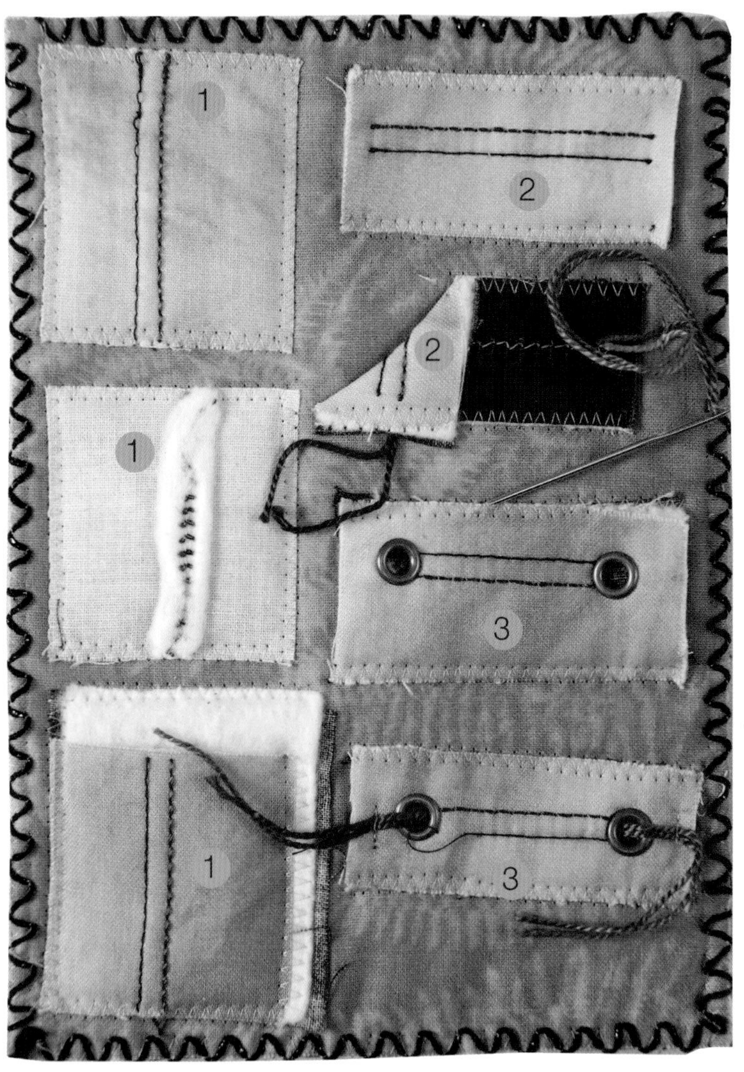

Edge finish: repeated zigzag stitching

1 FAUX CORDED TRAPUNTO:
Layer the first layer of batting and fabric. Baste with a wide twin needle. Trim the excess batting from the outside edges, leaving a twin stitched band of batting. Add a second batting. Fill stitch to either side.

2 TWIN NEEDLE CHANNEL CORDED TRAPUNTO:
To cord from the back, twin needle stitch through fabric, batting, and backing. Using either a large eye craft needle or a bodkin, insert yarn or cording into the twin needle channel from the back or through eyelets on the front. Draw the cord through the length of the channel. Secure the cord in place. Stitch closed any hole caused by the bodkin.

3 TWIN NEEDLE CORDED TRAPUNTO:
To cord from the front, set eyelets, grommets, or buttonholes at entry and exit points through fabric and batting. (Not backing, unless you want a hole completely through the project.) Twin needle stitch between these points. Use a large eye craft needle or bodkin to draw the cord through the channel and secure.

Designer's Workshop

Here's another example of using a closed shape for a trapunto effect. We stitched around the heart and then stippled in a meander stitch.

stamped & corded trapunto

This trapunto passport page includes trapunto with a stamped image as well as corded trapunto. Embroidery thread was hand stitched over the cords to accentuate the trapunto effect. Refer to Rubber Stamping directions on page 170.

Edge finish: couching with yarn and pearl cotton

1 Stamp fabric with a stamp of your choice. Place a thick batting under the fabric. Stitch around the outline of the shapes in your stamp image that you would like to trapunto. Twin needle stitch a 5" parallel line to either side of the stamp. Stitch two eyelets for the start and finish of the corded trapunto. Trim the first batting layer around the eyelets, diamonds and twin needle lines.

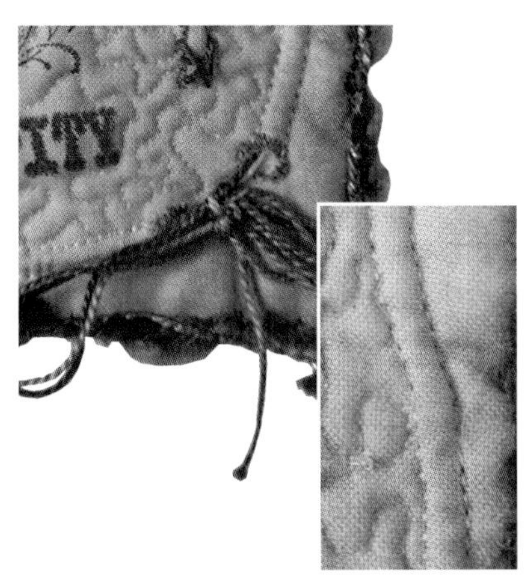

2 Layer the second batting, and backing. Add the corded trapunto by using a wide-spaced twin needle, beginning and ending the channel at the eyelets, stitching in an irregular rectangle shape.

3 Add fill stitching around the stamped motifs and corded lines. Hand embellish the two corded lines with x's, using a hand-dyed cotton or thread of your choice.

4 Using a large eye needle, work a double strand of pearl cotton through the stitched rectangle channel. Tie the pearl cotton tails in the lower right corner into a bow.

sewing with ultra heavy threads

Ultra heavy threads are too thick to travel through the upper tension disks on our machines. To solve this dilemma we use ultra heavy threads in the bobbin and bypass the upper tension system. This means we have to sew upside down which can have surprising results.

Edge finish: zigzag bobbin stitched with ultra heavy metallic thread

getting started

Learning how to control and work with ultra heavy threads to their best advantage keeps those surprise results happy ones. The tension of the bobbin case allows a single stitch length's thread to pull into a stitch. If this tension is too tight it will shorten the lower portion of the stitch and pull the top thread down to the back of the fabric. If it is too loose, excess bobbin thread could pull to the top of the fabric. A slight amount of tension from either a specialized bobbin case or loosened bobbin tension will often offer just enough control of the ultra heavy thread. Refer to Bobbin Case information on page 16.

Machine set up:
Normal, bobbin sewing, and free motion

Tension:
Balanced
Adjust as needed for bobbin sewing

Needle:
Size based on top thread

Top thread:
Fine and medium weight

Bobbin:
Ultra heavy and fine

Bobbin case:
A special bobbin sewing case, if you have one, or the ability to loosen or bypass bobbin tension

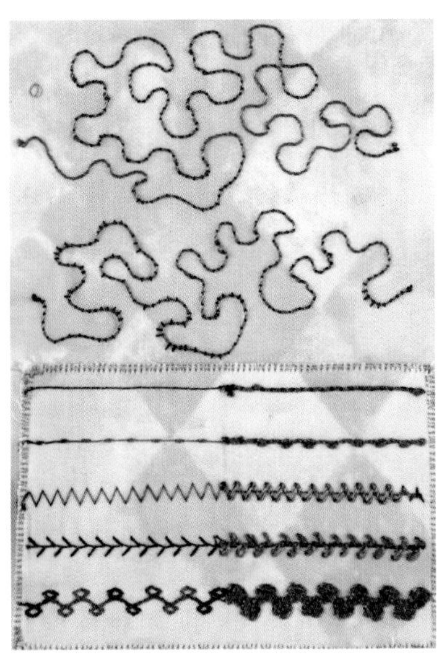

Try these starter steps on a Passport page or separate practice fabric.

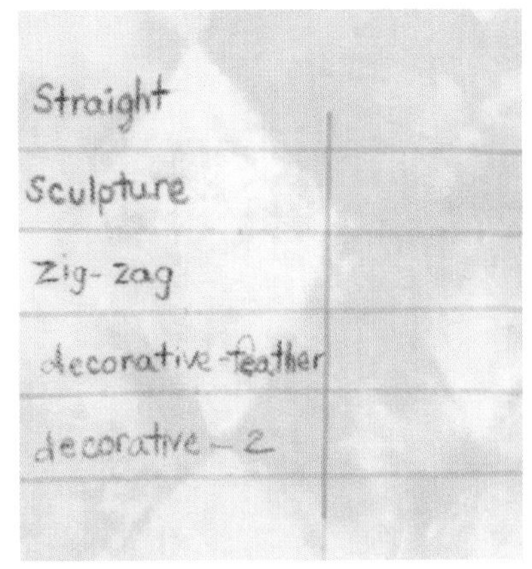

1 Make a grid by drawing a 4" x 2-1/2" rectangle on the lower portion of the fabric base page. Add five horizontal lines and one line down the center. Stitch across the five lines with the fine thread. This initial stitch line acts as a marking line for bobbin stitching.

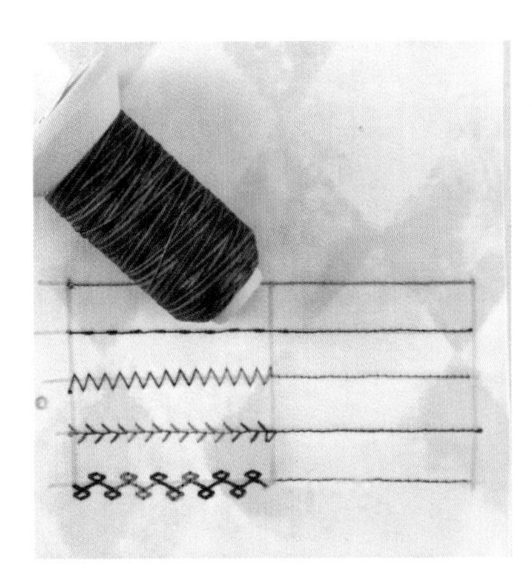

2 Choose a variety of stitches to show the heavy thread in different situations. Stitch across the first half of the grid (left side) with medium weight thread to show the stitches in normal conditions. Be careful not to continue across the center line.

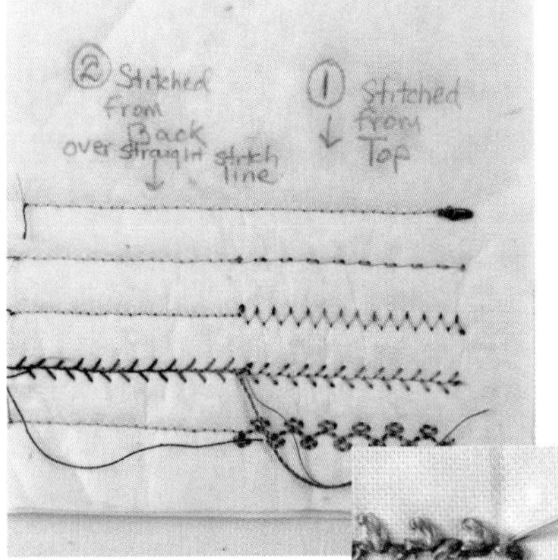

3 Loosen the bobbin case tension to accommodate your ultra heavy thread. Turn the fabric over. Following the previously stitched marking line, begin at the center and draw up the bobbin thread or leave a long enough tail to keep it out of the stitch line. Stitch on the marked line and finish with a long tail. Draw the tails to the reverse side, knot, and trim. Repeat this using several different stitches.

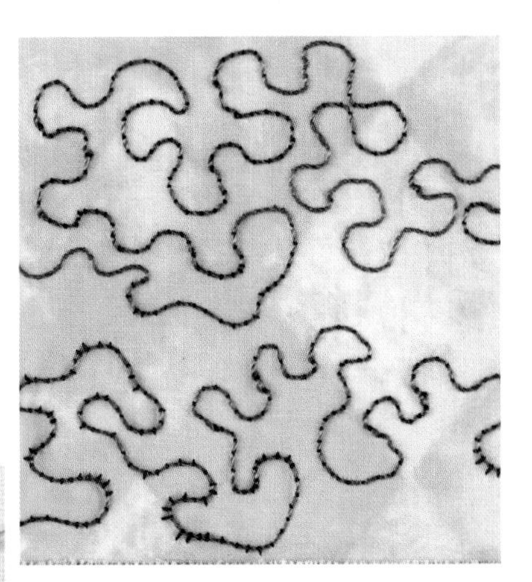

4 Set the machine up for free motion stitching with ultra heavy thread in the bobbin. Sew with a medium thread in a blending color on top. Repeat with a contrasting color. Tie off the thread tails. Notice the difference the top thread makes.

bobbin thread party

You know when these threads are in the room. Even though they all fall into the ultra heavy category, there are distinct differences between them. Ribbons, yarns, and floss can all become thread when applied by machine from the position of the bobbin thread.

Edge finish: zigzag couched chenille yarn.

1 Draw pennant shapes on the back of pre-fused silk fabric. Trim the shapes with a pinking shears and fuse in place across the two pages. Stitch in place to secure. Stamp or write the letters B-O-B-B-I-N on the fabric scraps. Cut out the squares, approximately 3/4", and stitch in place with a few rounds of straight stitching.

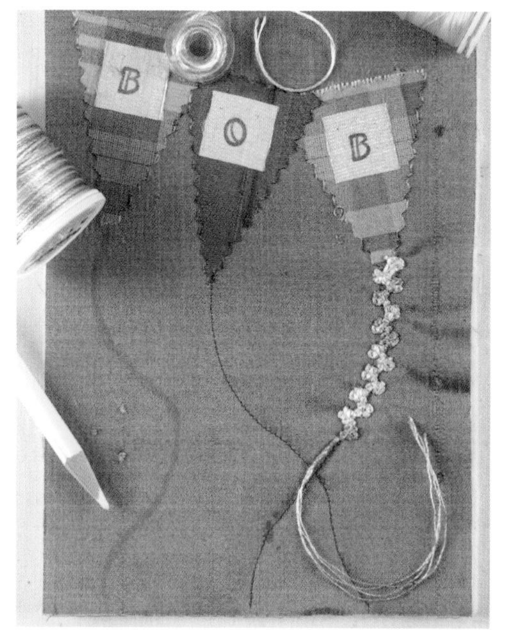

2 Use your favorite marking tool to designate the tail lines on the front of the fabric. Stitch the marked line with a fine thread. Turn the fabric over and stitch the assorted bobbin thread tails from the back following the pre-stitched lines. Be sure to leave long tails to draw to the back and tie off. Repeat the same process for stitching the ribbon across the pennant tops. If desired, add little bobbin-stitched tails at either end of this line.

tip

- Winding ultra heavy threads and fibers onto the bobbin can be tricky. Your machine's bobbin winder may not always handle the job. In these cases, or when using a fiber not suitable for machine winding, wind the bobbin by hand. Keep a moderate amount of tension on the fiber, but don't pull too tightly.

ultra heavy free motion

Machine set up:
Free motion
drop feed dogs

Tension:
Slightly tightened top.
Set up for bobbin sewing

Needle:
Embroidery 90/14

Threads:
Assorted decorative

Edge finish: zigzag bobbin stitched with ultra heavy thread

1 Layer your base fabric with a piece of medium to heavy weight stabilizer. Choose a top thread that you won't mind seeing peek through. Stitch a free motion meander bobbin stitch pattern over the base fabric. Light to moderate bobbin tension gives these stitches a tighter appearance.

2 Choose a fabric for your free stitched squares. We used upholstery weight fabric but others work well also. Add a stabilizer if using lighter weight fabrics. Cut the fabric approximately 3" x 4". Bobbin stitch a grid of wavy lines in straight stitches from an assortment of ultra heavy threads. Cut out two 1-1/2" squares and set aside as you stitch the flower bobble.

Making the Flower Bobble

Note: The unique organic look of this flower is achieved by completely bypassing the bobbin tension system. If the tension cannot be bypassed, similar results can be achieved by fully opening the bobbin tension and lowering the upper tension.

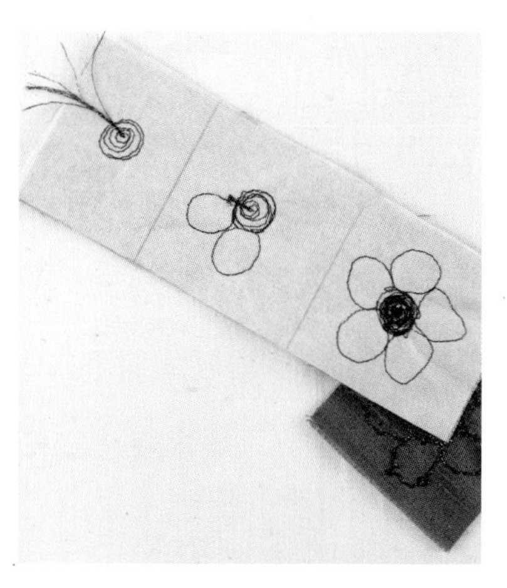

3 Draw the bobbin thread up and pull it to the side to be clipped after it is crossed by a few stitches. Lightly stitch the inner center of the flower, then move out to create five to six petals.

4 When the petals are complete, move back to the center. Now the bobble starts to form. Work from the outer edge toward the center with the goal of finishing up in the center. This will be the peak of the bobble. You will feel the fabric rise from the bed of the machine with the bulk of the thread. The needle will start to protest noisily and you will feel when it is time to stop stitching.

5 Snip the bobbin thread close to the bobble.

6 Arrange your flower bobble and your small squares on your fabric base and stitch in place.

For a different look

Free stitch with a straight or zigzag stitch in a spiral direction. Repeat with one or more threads. Cut the resulting shape into smaller pieces and appliqué them as desired.

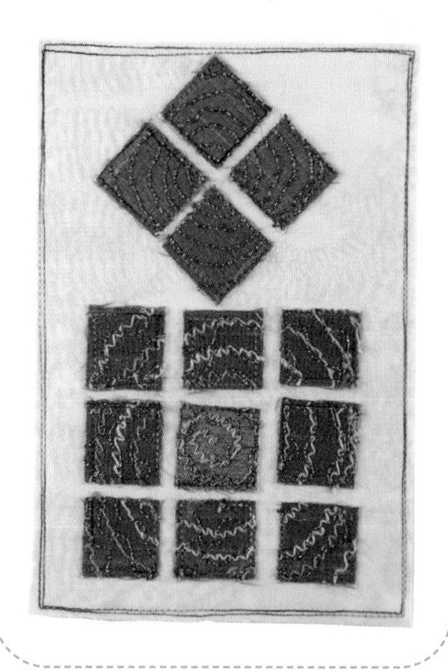

hand look quilt stitch by machine

The hand look quilt stitch, also known as the mock quilt stitch or sculpture stitch, is a versatile addition to your stitch repertoire. It gives the look of hand quilting and can be used as an option for sashiko inspired stitching. The trick to this stitch is sewing with an invisible or very fine blending thread in the top while the thread you want to show is in the bobbin. The bobbin thread is pulled to the top by setting your sewing machine tension to 9 or its maximum setting. Results will vary depending on needle, batting, fabric choice, and the weight of thread in the bobbin. Larger needles will cause a bigger hole in the fabric to allow for pulling the thread up from the bobbin. Heavier threads from the bobbin can make a bolder stitch statement, but tend to be less consistent in pulling through to the top. Invisible thread on the top can shine more than a finer, blending thread, but more consistently pulls the bobbin thread to the top. A longer stitch gives a more primitive look to the stitching, while shorter stitches look more traditional. Try a variety of threads to discover the best combination for you and your machine.

Machine set up:
Normal

Tension:
Top at 9
Bobbin normal to slightly loose

Needle:
Embroidery 90/14 OR Topstitch 100/16

Top thread:
Invisible and assortment of ultra fine

Bobbin:
Fine, medium, and heavy weight in a variety of polyester, cotton or rayon

Pre-wind a variety of bobbins in assorted threads. A walking foot gives better control for this stitching, but use what works best for your machine set-up.

Draw a 2" x 3" rectangle in the center, upper portion of the background fabric. Layer the fabric over batting, not a stabilizer. Stabilizers could inhibit the forming of the stitch.

Stitch using different combinations of threads and stitch lengths. Stitch the first combination on the marked frame. Stitch concentric frames using the sewing machine foot as a guide for the width between frames.

Edge finish: heavy thread zigzag stitch

Once you have tried several combinations, finish the page by stitching around the perimeter in the same stitch, but with medium thread at a normal tension from the top. This will demonstrate the "normal" version of the stitch. Add the hand written message "Hand-look quilt stitch" with a sketch of the stitch pattern, if desired. This will serve as a reminder when you look to your Passport for reference. Make a note of your favorite thread combinations and settings. Add a few buttons to finish the page.

Heavy Thread Collage

Three different fabrics are stitched with an assortment of ultra heavy and medium weight threads. They are then cut into smaller pieces, tacked in place, and the seams are covered with machine wrapped cords.

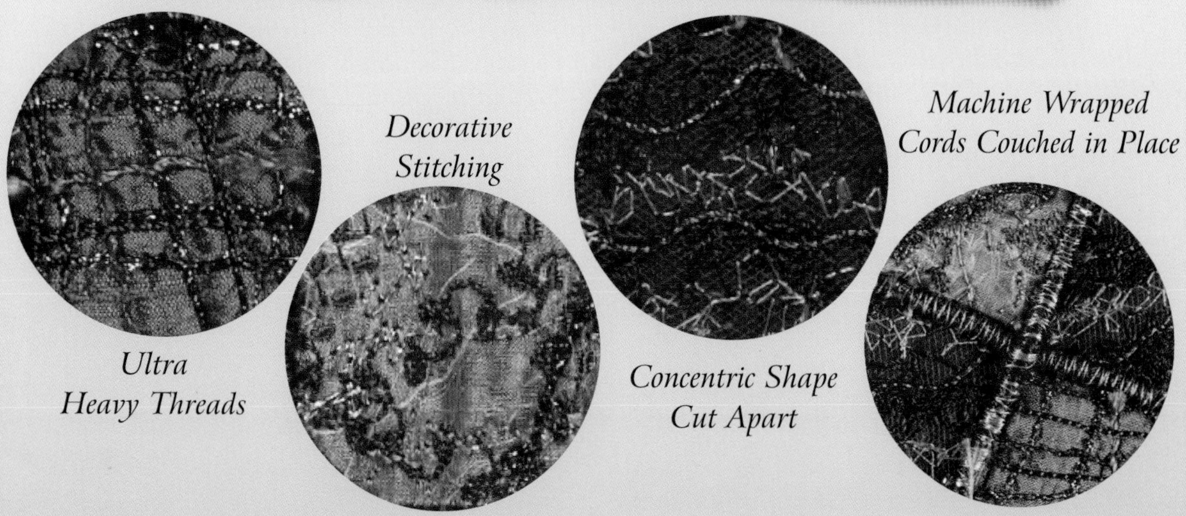

the techniques

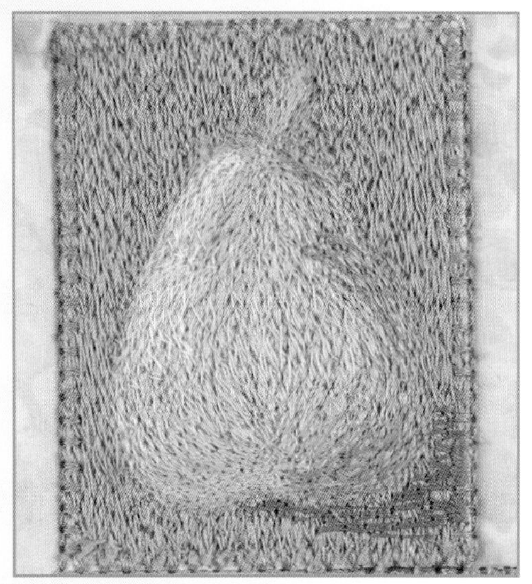
thread painting

thread painting on printed fabrics

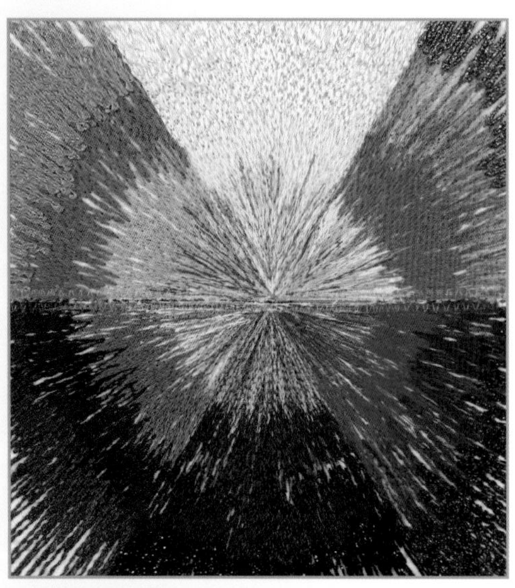
stitched color wheel

thread blending

drawing & sketching

thread painted appliqué

Thread as Paint

Thread has been used as paint on fabric since the early Middle Ages. We suspect the reasons then are the same as now; beauty, expression of story, and just plain fun. Thread sketching and painting are also called free motion embroidery but we feel sketching and painting better describe the free flowing build up of stitches.

Thread painting is incredibly versatile. Thread colors can be layered and blended to create lush and intricate pieces. Add just a few stitch details to a fabric motif or completely cover the fabric in thread.

Thread sketching is simply using your needle to sketch an image or shape. Draw a still life, portrait, landscape, or anything that captures your interest.

The exercises in this section will give you confidence to let the thread be your paint and the needle your brush.

thread painting

Dense thread painting may look intricate and difficult but it is as easy as coloring with crayons once you know a few tricks. The most difficult aspect is controlling the distortion of the fabric, which is why we stitched on a heavy stabilizer. We stitched pears using a straight free motion stitch, a zigzag free motion stitch, and a circular stitch which give lots of texture but require an underlying stabilizing grid. You will learn how the direction of the stitching affects the base fabric.

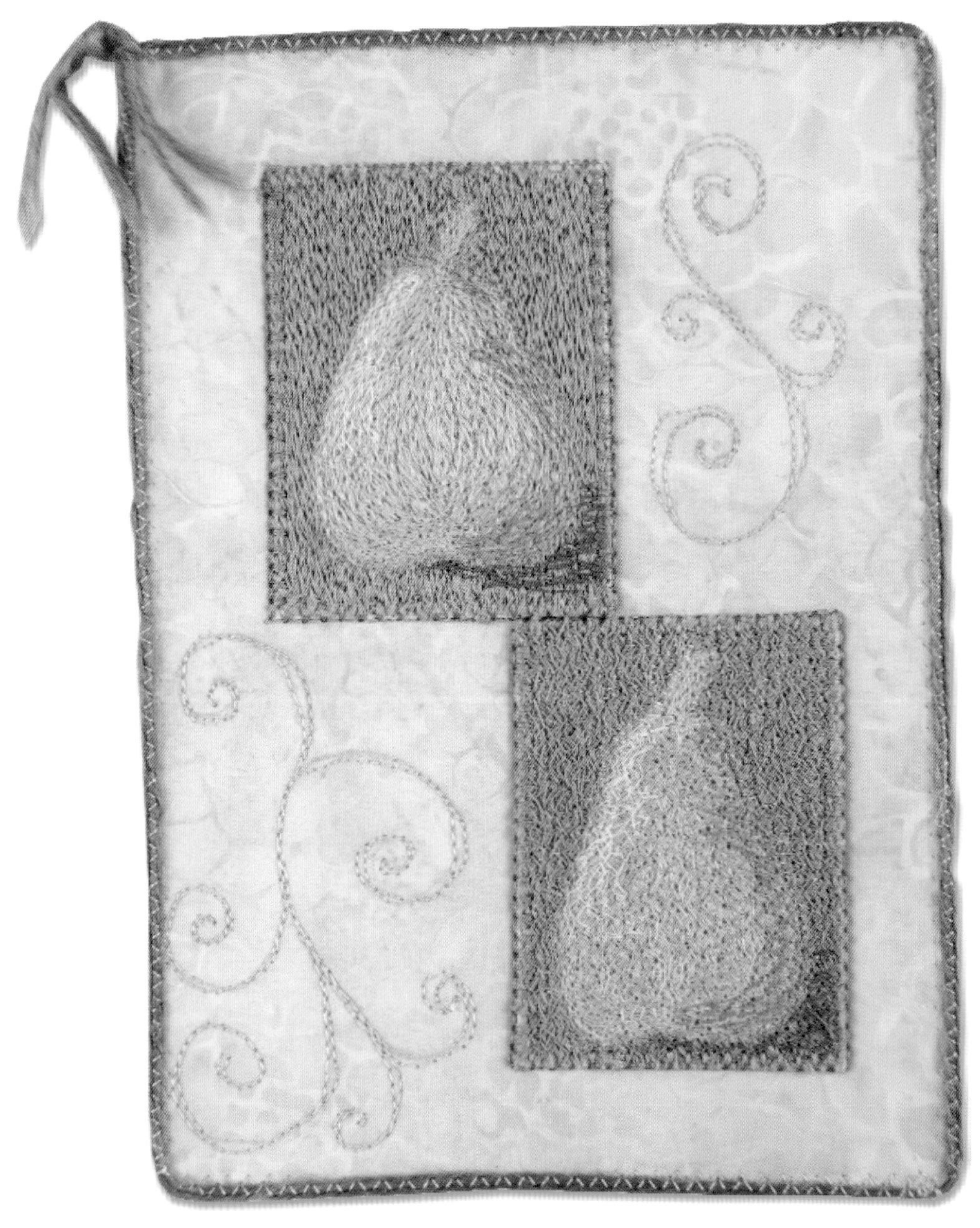

Edge finish: couched yarn with zigzag stitch.

Machine set up:
Free motion
drop feed dogs

Tension:
Slightly loosen upper;
top thread should be pulled
slightly to the back of your work

Needle:
Embroidery 90/14

Top thread:
Any decorative;
try a medium weight to begin

Bobbin:
Ultra fine or fine polyester

Straight Stitch Pear

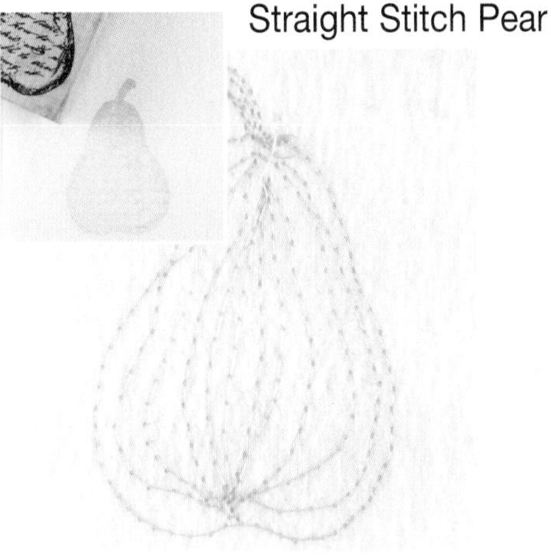

1 Draw a pear or use a rubber stamp to create a pear image on a heavy stabilizer. Stitch around the outline of the pear shape. Lightly stitch the interior following the contour of the pear shape.

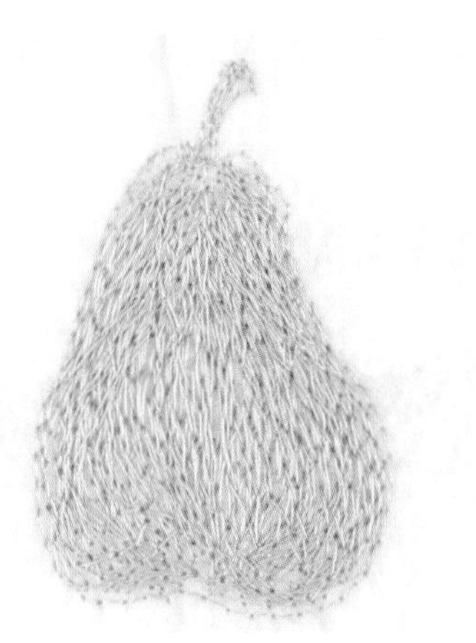

2 Fill the interior of the pear by continuing to follow its contour lines. Notice the stabilizer is becoming distorted. Press the pear and continue to fill in the shape.

3 Press again and draw a box around the pear.

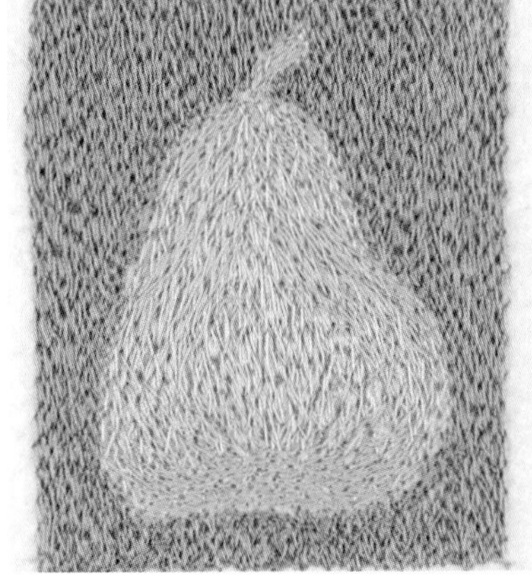

4 Fill in the box area with a background color. Press. Notice the distortion of the stabilizer. This dense stitching is very difficult to do on a tightly woven fabric.

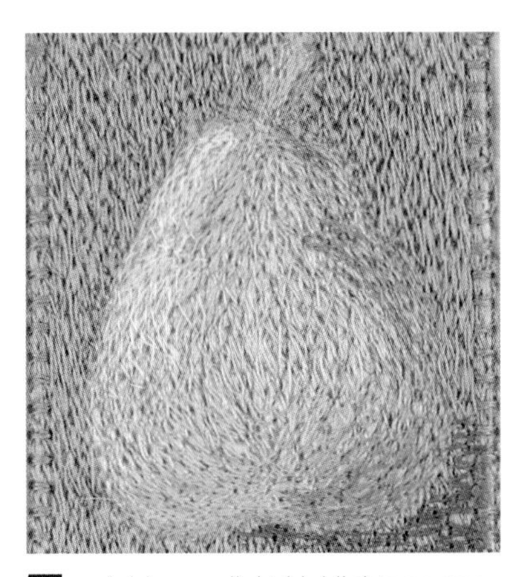

5 Add some light highlights on one side of the pear using a very pale version of the pear color. Add some shadow on the opposite side using a darker shade of the pear color. Add some shadow to the background just under and to the side of the pear. Press. Place a piece of fusible web on the back of the rectangle and fuse in place. Cut out the pear rectangle and fuse it to your fabric page base.

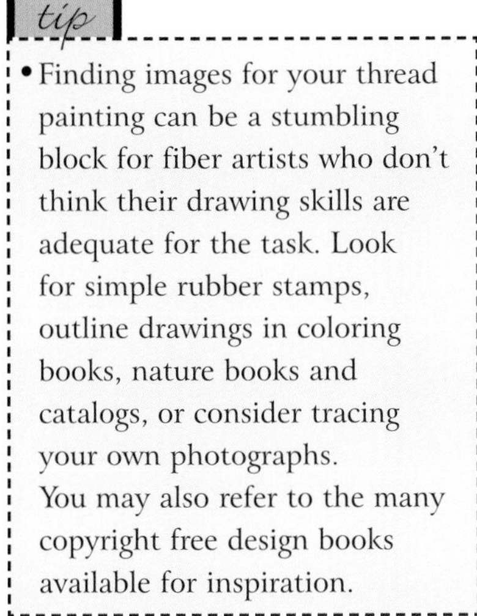

tip

- Finding images for your thread painting can be a stumbling block for fiber artists who don't think their drawing skills are adequate for the task. Look for simple rubber stamps, outline drawings in coloring books, nature books and catalogs, or consider tracing your own photographs.
You may also refer to the many copyright free design books available for inspiration.

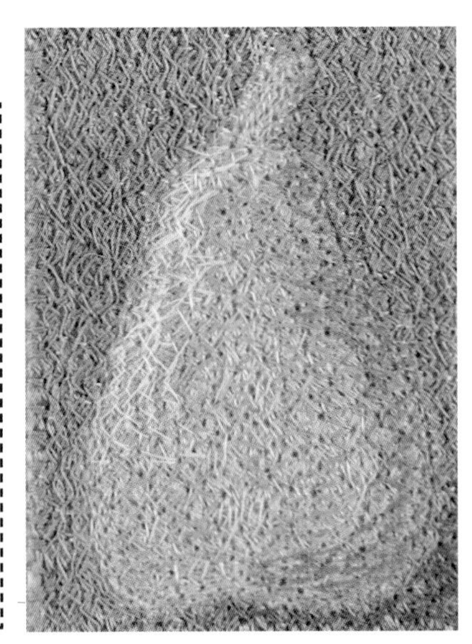

Zigzag Stitch Pear

Repeat the steps for the straight stitch pear using a zigzag stitch. Notice how different the pear looks. The zigzag stitch gives a casual, energetic look to the pear and minimizes the distortion of the stabilizer.

stitching pears

For this pear we used a variegated hand-dyed green thread. For extra texture we used a twisted two-color rayon thread for the background.

Circular Stitch Pear

Edge finish: ribbon with zigzag stitch.

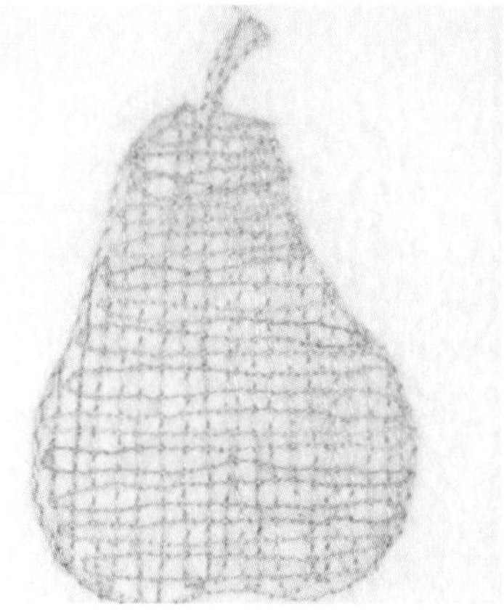

1 Draw a pear or use a rubber stamp to create a pear image on heavy stabilizer. The circular granite stitch tends to distort woven fabric a great deal, so begin by stitching an underlying grid inside the pear shape.

2 Stitch around the exterior of the pear in small circles. Follow steps 2-5 from the straight stitch pear on page 69 but substitute circular stitching for the straight stitching. Fill in the interior of the pear.

To control fabric distortion

- Use a light weight (60, 80 or 100wt) thread in the bobbin. The bobbin thread will last longer, won't add bulk, and will decrease fabric distortion.
- Tear away stabilizer can be layered to offer extra support.
- Stitching over the entire design as a whole rather than stitching one section at a time will help minimize distortion.
- Press often with a steam iron.
- Choose a stabilizer as your base. Tightly woven fabrics such as batiks are very difficult to use successfully. You may want to experiment with an open weave fabric.

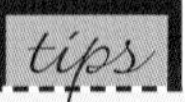

- In thread painting the length of the stitch is controlled by your machine speed and the speed you move the fabric. Experiment with the different looks of small stitches and long stitches, heavy threads, and light weight threads.
- Loosen the top tension so the top thread is pulled slightly to the back of your work. This will ensure dots of bobbin thread won't appear on the top of your project.
- Relax and don't worry if you color outside the lines or if your stitching line goes astray. You can incorporate these wayward threads with additional stitching or remove them later.
- Sew more slowly when working on smaller or more intricate designs.
- Use a temporary spray adhesive as an alternative to pins when securing the stabilizer to the fabric while stitching.
- When thread painting directly on fabric use a machine embroidery hoop to stabilize the fabric and avoid hand fatigue. Hold the hoop lightly with the fingers on the outside edges rather than gripping it.
- Use a fusible web on the back of a heavy, thread painted piece before you trim it to size to add to your final work. This will help secure the threads when you trim.

Designer's Workshop

Pull elements from a large scale pattern to embellish with thread painting. The apron pocket gets texture from heat shrinking threads.

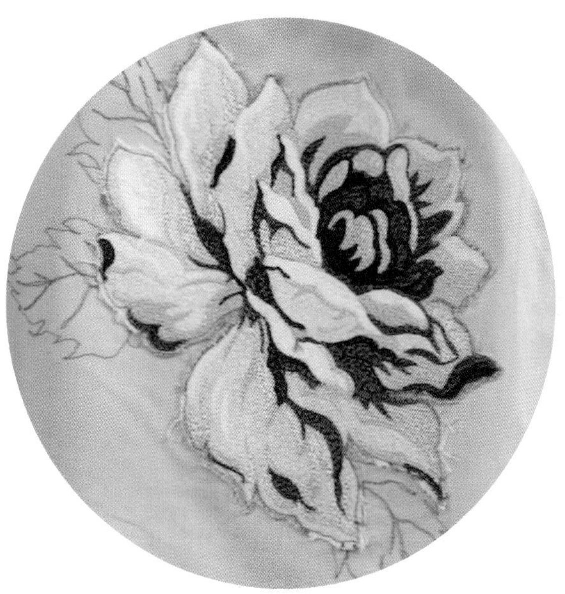

Thread Painting

thread painting on printed fabrics

A great way to start thread painting is to use photographs or printed fabric as a coloring book. The design is already there to embellish with thread. You can use either a commercial fabric or a photo you have printed on fabric. Feel free to color outside the lines, change the design and color, and play and doodle as you stitch. Try stitching with blending and contrasting threads.

Edge finish: couched yarn with zigzag stitch.

Machine set up:
Free motion
drop feed dogs

Tension:
Loosen upper

Needle:
Embroidery 90/14

Top thread:
Decorative embroidery or cotton

Bobbin:
Ultra light weight or light weight

1 Print a photo on fabric prepared for an inkjet printer.

2 Cut a piece of medium or heavy weight stabilizer and secure it to the back of the fabric. Outline stitch the element you want to embellish with thread.

3 Lightly stitch, working around the entire motif. Follow the contours and shape of the motif. Press.

4 Continue to fill in the motif until you are pleased with the stitching. Press and trim to desired size.

tips

- Printed fabrics prepared for inkjet printing usually have fairly high thread counts so you will not be able to completely fill in the motif. Stitch only until the fabric begins to distort.
- Press often to help control distortion.
- Stitching evenly around the entire motif also helps to control distortion.
- If you get carried away with stitching and end up with too much distortion simply cut out your image and use it as an appliqué.

Doodling on Printed Fabric

1. Choose a fabric with a large graphic design. Back it with a medium weight stabilizer.
2. Doodle stitch on top of the design. Don't limit yourself to matching thread colors or coloring in the shapes. Play and experiment, outline motifs, color them in, and add new designs or shapes.

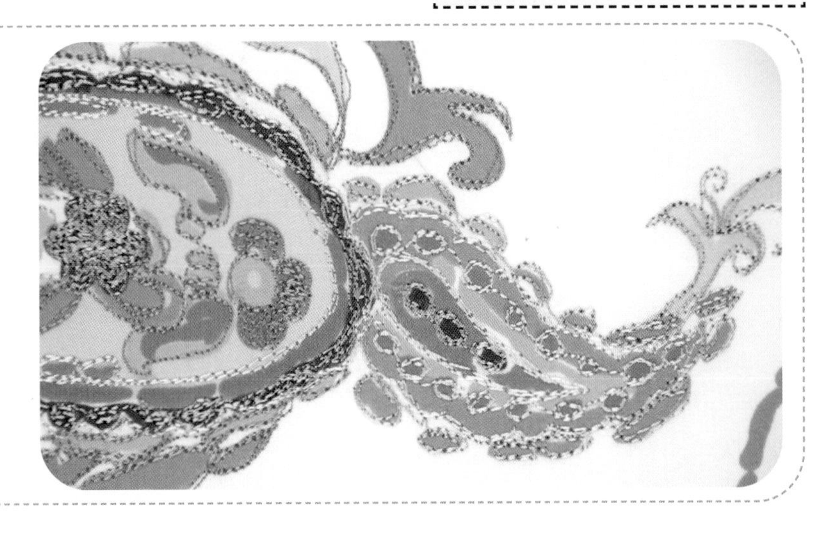

stitched portraits

Stitching a portrait can be intimidating. These simplified solutions boost confidence and show there's more than one way to work with photos.

Edge finish: couched yarn with zigzag stitch.

getting started

There are many ready-to-print fabric products on the market. They range from silk to heavy cotton duck. To keep things simple, we used two ready-to-print paper-backed fabrics. Follow the manufacturer's directions and consult your printer manual before using. Look at your photo to find lines to highlight with stitching. Once you establish your stitch lines, decide what and how much stitching you want in the background and surrounding fabrics.

The Classic Grid

Dividing the elements of a composition is one way to instantly add artistic flair. It involves sewing concentric squares around the pieces with a walking foot.

Pre-fuse the print with fusible web.

Trim your printed image as needed to maintain a square shape. Divide the image and cut.

Lay the cut pieces on a fabric backed with stabilizer and fuse in place.

Stitch around each piece multiple times, highlighting a focal piece if desired. Next, stitch around the whole picture. Add appliquéd letters if desired. Add backing fabric and finish the edges.

Sheer Image

Printed silk organza adds a new level of interest to work.

Print the photo image on the silk organza. Remove and save the backing paper.

Create a background with light and dark fabrics and stitching. Frame the image with a straight stitch in a fine thread. Stitch over the highlight lines with metallic thread.

Use the framing stitch line as a guide to bobbin stitch with ribbon.

Cotton Image

Print and fuse the photo image to the fabric background.

Decide how you will stitch the highlight lines as well as those that follow the image off the photo and onto the background. In this case, the shirt and arm lines are continued off the photo onto the background. Use your favorite quilting stitches to complete the piece.

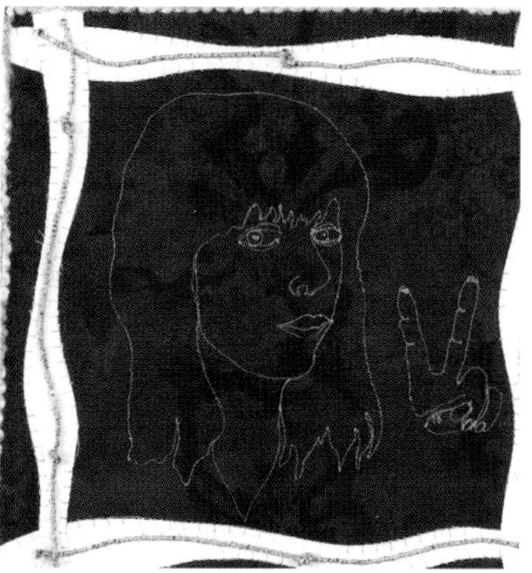

Bobbin Stitching

The backing paper saved from the silk organza provides an image to use as a guide for sewing from the bobbin. Test your heavy weight threads by stitching on scrap fabric.

Place the stabilizer on the back of the fabric. Spray-baste or pin the paper image over the stabilizer. Stitch the highlight lines.

Turn the image over whenever threads need to be snipped or you need to double-check the stitching.

Frame the image with fused wavy fabric strips and machine wrapped cords.

- Use stabilizer. Refer to Stabilizers on page 19.
- Do not over stitch the printed image as it could cause distortion or image degeneration from too many stitches.
- Look for fairly close up photos with good contrast.

stitched color wheel

The color wheel is a dependable guide through all kinds of color-coordinating cataclysms. Stitching a color wheel of your own, in a variety of thread weights, will not only breed a familiarity with color, but will also help you build a resourceful tool.

Edge finish: couching with bulky yarn and multi-color thread

Try This!

Delve deeper into color exploration by creating a 12-step color wheel.

Edge finish: a zigzag stitch and a fine yarn couched around the page twice

getting started

When filling an area, do not work any small area too intensely. Work in waves across the area, passing back and forth three or more times to lessen distortion. Work in one section and then the opposite section to balance the tension in the background fabric. Work with slightly decreased tension to allow the upper threads to be pulled to the underside of the fabric to avoid the bobbin thread peeking to the top. Press from the back to tame the distortion that will increase with each circle. Much of this will release when the piece is cut into two separate Passport pages.

Machine set up:
Normal and free motion

Top threads:
Medium weight in a color that blends with the background fabric

Yellow, green, blue, purple, red and orange each in a medium, heavy and ultra heavy weight

Multi-color in medium weight

Bobbin threads:
Yellow, green, blue, purple, red and orange each in a medium weight

Medium weight in a color that blends with the background fabric

Needles:
Embroidery 90/14, Topstitch 16

NOTE: This double page spread calls for a lot of thread. If your collection does not quite support so many colors and weights, scale back. Complete a 6 or 12-color wheel in only one weight of thread instead.

1 Layer a 9" x 12" piece of muted fabric with a heavy stabilizer, fuse or pin together. Using a chalk or erasable marker, draw a 10" x 7" rectangle centered on the fabric page. Mark the intersection points by placing a mark along the side lines at 1/2" from the top and bottom, as well as a center point along the top and bottom lines at 5". Draw the center line and mark the center point at 3-1/2" to indicate the intersection point. Draw diagonal lines between the intersecting points. Stitch along the diagonal and center lines. These lines mark the sections of the 6-color wheel and secure the stabilizer in place. Do not stitch the outside lines of the rectangle.

2 Designate a color for each of the six wedges. Free motion stitch the inner circle using a straight stitch and a multi-color, medium weight thread featuring the color of the wedge. Stitch in a forward and reverse motion, radiating out from the center point. DO NOT completely fill the area or it will loose some of its sense of radiance. The gradual filling of the area is shown across three colors in the photo.

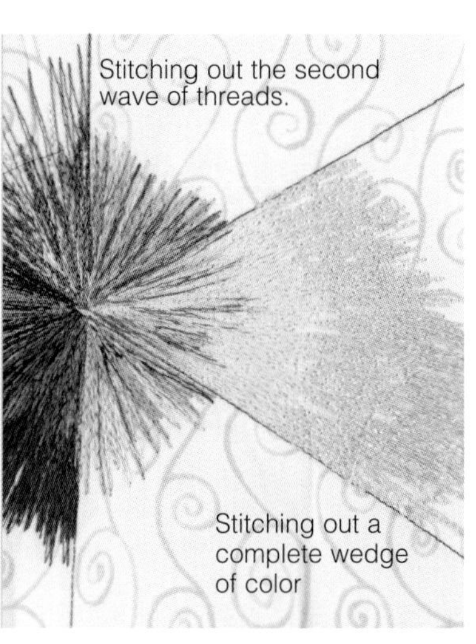

3 Stitch the next circle in medium weight cotton threads. Follow the same process as in step 2.

4 As you switch to heavy threads for the third circle, you may need to decrease the upper tension. You will notice these sections require less stitching than the medium weight threads because the thread fills the space faster. Follow the same stitching process as in step 2.

5 The final circle is stitched with ultra heavy threads in the bobbin. Refer to Sewing with Ultra Heavy Threads on page 58 for tips on this technique. Use a matching medium weight thread on top. Unlike the other sections, you will be able to fill the ultra heavy section in one pass with minimal distortion. The outer edge of the circle is kept fairly open to create the feeling of radiating rays of color.

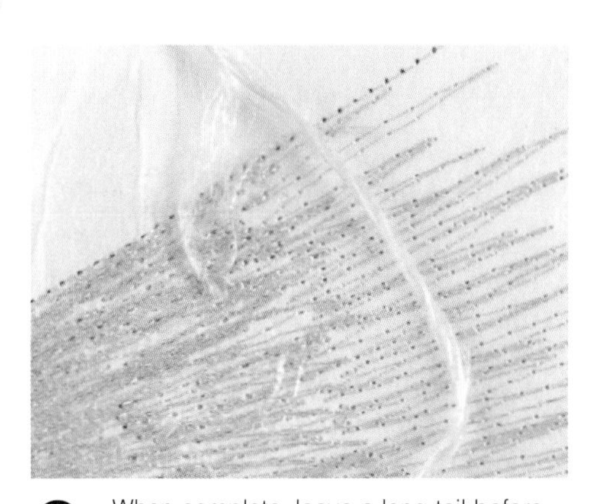

6 When complete, leave a long tail before cutting threads. Draw this tail to the back and secure with a knot or fuse to the back with a piece of scrap fabric. Press the page and carefully trim it in half. Trim each half of the page to 5" x 7". Be careful to keep the center point of the long, inside edge of each page at the 3-1/2" mark. Embellish only the outside edges of this pair of pages so as not to detract from the color wheel effect. Finish the abutting inside page edges with a zigzag of clear monofilament thread to prevent unraveling.

underpainting

Underpainting is a technique used for centuries in the fine art world to help define color values for later painting. Thread artists can also use this technique. Underpainting can add interesting color play and subtle variation.

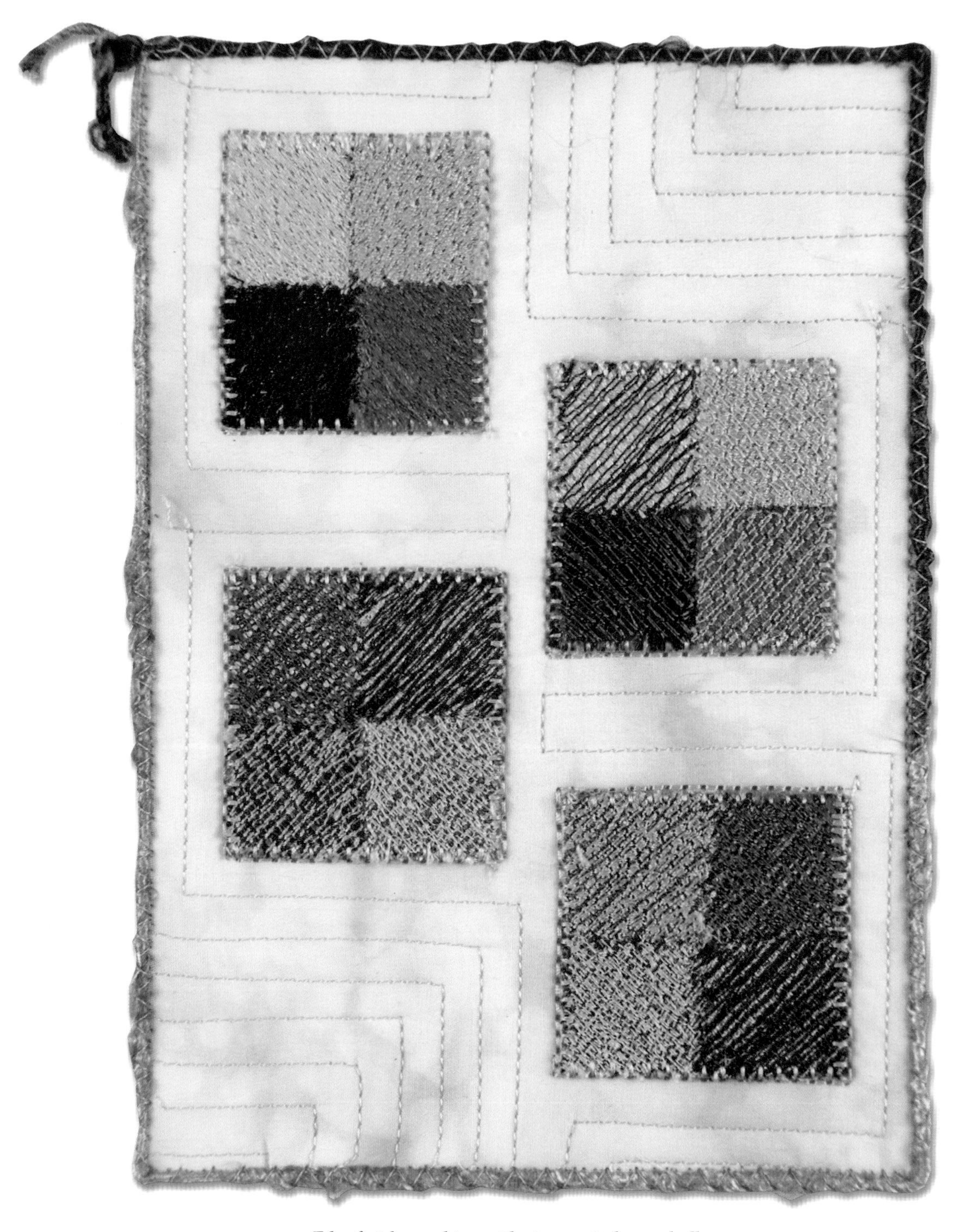

Edge finish: couching with zigzag stitch over bulky yarn

getting started

What happens when you stitch green over blue or violet over orange? You will see that each layer of thread adds a complexity and interest that enriches your stitching.

Machine set up:
Free motion
drop feed dogs

Needle:
Embroidery 90/14

Tension:
Loosen top slightly

Top thread:
Decorative cotton, polyester, silk or rayon

Bobbin:
Light weight or matching

1 Cut a 6" x 12" piece of heavy stabilizer. Draw four evenly spaced 2" squares on the stabilizer. Divide each square into four 1" sections. Label the 2" squares 1-4.

2 Choose threads in two warm colors and two cool colors. Choose threads similar in value. Stitch each 1" section with a different color thread. Our sample squares are stitched at an angle from the middle out to the opposite corner. You can stitch on the diagonal, vertical, or horizontal but keep the angle the same in each square. Do this on all four stabilizer squares as your base layer. Square 1 is finished.

3 The next layer of stitching will be done on squares 2-4. Change the direction of your stitching. If you stitched vertically on the first layer, stitch horizontally on the second. This allows the thread colors to lie on top of each other as a separate layer. Next rotate the thread colors clockwise one square. Stitch this on all three squares. Square 2 is finished after this step.

4 The next layer of stitching will be done on squares 3-4. Change the stitch direction again. Stitch in the same direction as your first layer of thread in step 2. Rotate the thread colors clockwise one square again. Stitch on both squares. Square 3 is finished after this step.

5 The final layer of stitching is done on square 4. Stitch in the same direction as in step 3. Rotate the thread colors clockwise one more square. This final square is done.

6 Apply fusible web to the back of the stabilizer and cut out the four squares. Fuse the squares to your page base fabric. Secure with a machine blanket stitch if desired. Quilt page background if desired.

thread blending

There are times when you want to blend multiple thread colors for subtle shading or rich more complex color. Blending differs from underpainting as the stitches lay parallel to each other rather than perpendicular.

Edge finish: zigzag stitching with variegated polyester thread

getting started

An easy way to blend thread colors is to use two threads through the needle at one time. Our exercises show blending with light and darker tones as well as with colors adjacent or analogous on the color wheel.

Machine set up:
Free motion
drop feed dogs

Needle:
Embroidery 90/14

Tension:
Loosen top slightly

Top thread:
Decorative cotton, polyester, silk or rayon. Refer to sewing machine manual for threading two threads through the needle.

Bobbin:
Light weight or thread to match top

1 To make a blended heart choose three values of one thread color. We used red violet in a light, medium, and dark value. Draw the blending heart pattern below on a piece of heavy stabilizer. Choose the medium value color thread and thread paint the inside smaller heart. Refer to Thread Painting on page 66 before beginning.

2 Thread the needle with the light value color thread and the medium value color thread. Thread paint the left side of the larger heart. If desired, stitch a defining line around the inside heart with the light thread color.

3 Thread the needle with the medium value color thread and the dark value color thread. Thread paint the right side of the large heart. If desired, stitch a defining line around the inside heart with the dark color. Apply fusible web to the back of the finished heart and cut out. Fuse to your passport page.

To stitch the analogous stitched heart on the bottom of the Passport page choose three thread colors that lie next to each other on the color wheel and follow the steps above. We used a violet, a red violet, and a light red thread.

Blending Heart Pattern

drawing & sketching

As with pencil sketching, thread sketching is about observation. The more we draw or sketch, the more we learn to look closely and see form, light and texture and the better our skills become. Sketching is not photo-realism. A sketch strives to capture the essence of something rather than be an exact replica of it.

Edge finish: zigzag stitching with a ribbon tacked in place with cross-stitches.

getting started

Let your thread give shape to objects with straight, curved, and jagged stitch lines. These are simple representations of some common sketching terms:

Jagged Straight Stitch

Contour

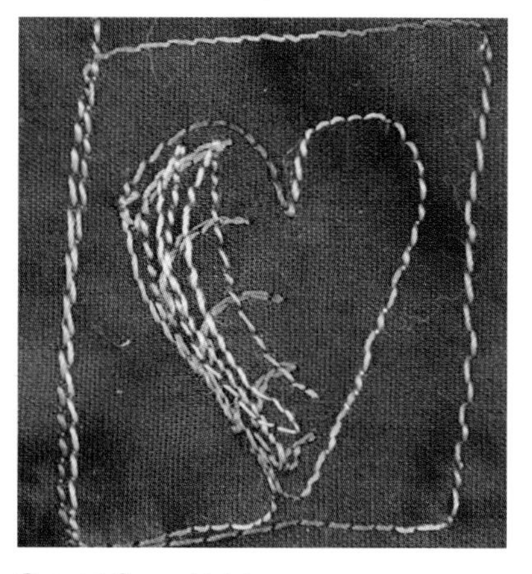
Curved Cross Hatch

Cross Hatch

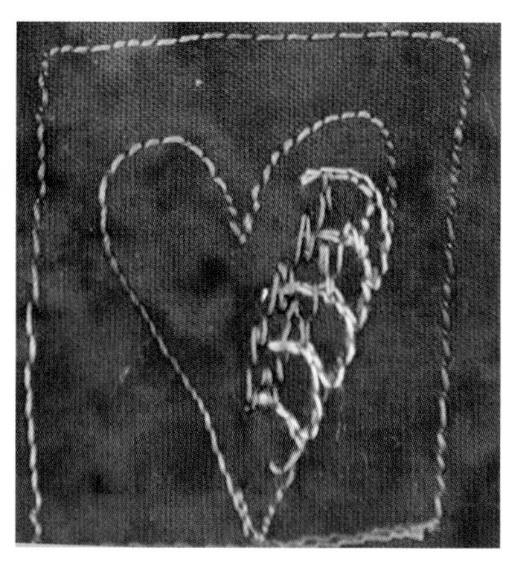
Jagged Contour

Shading

- Begin with free motion, straight stitching. Use contour lines and cross-hatching for shaping. Graduate to using jagged stitches, loops, and zigzags as you gain confidence.
- A thread sketch is lightly filled. The fabric should show through.
- Take time to step back and look at your sketch to get a fresh perspective.
- A sketched line is more forgiving than a single line drawing. If you have a wobble, blend it into the sketch with multiple lines of stitching. Don't worry about them. They add to the charm of a sketch.
- Stitch like you drive, looking ahead rather than at the needle.
- Press the thread sketch from the back if it begins to distort.
- Nothing says you can't change the orientation of your fabric. A line or shape may be easier to stitch from a different angle. If you are zigzag stitching, watch for how this affects stitch direction.
- Keep an eye out for "pause points." You can sense when you need to adjust your grip on your fabric or change your seating position. Watch for a natural point to stop with the needle down to re-adjust your position.
- An embroidery hoop can be helpful if your design is small.

drawing & sketching

To get you started in this technique we have included an outline drawing of a cup for you to trace. Or, you can draw your own image.

Machine set up:
Free motion
drop feed dogs

Tension:
Balanced

Needle:
Embroidery 90/14

Top Thread:
Any medium weight thread

Bobbin Thread:
Fine or medium

1 Trace an image onto tissue or plain paper. Use transfer paper to transfer the line drawing to a piece of fabric. Layer fabric with batting, if desired, and stabilizer.

2 Begin by stitching the outline. When possible stitch in a continuous line.

3 Continue stitching, travelling along the same lines to stitch different areas. Add touches of lighter and darker threads for contrast.

How many ways can you stitch a cup?

Abundant thread choices could fill a Passport.

from tracing to sketching

Tracing is a precursor to drawing. It helps train the eye and hand in capturing a representation of what you see. Try tracing and stitching the image of your hand and then move on to free sketching with these steps:

1. Lightly trace your hand onto background fabric. This may be slightly larger than other Passport pages.
2. Stitch over the outline in one continuous line to get the basic shape of the hand.
3. Restitch over the outline, this time dipping into the hand shape to create knuckles, fingernails, and wrinkles.
4. Add final details on top of the hand.
5. After stitching the traced hand try other thread sketches such as the bird shape on page 85.
6. Practice drawing on paper first. Transfer some sketch lines to your fabric and fill in with detail stitching.

Doodling

Edge finish: zigzag stitching with a ribbon tacked in place with cross-stitches.

1. Loosen up your stitching by doodling over printed fabric designs. Then take your doodling to the next level and doodle on blank fabric. Have fun and play.
2. Begin with a large scrap of fabric layered over a medium weight stabilizer. Depress the foot pedal and move your fabric under the needle as you stitch. Doodle and draw on your scrap fabric until you start to feel comfortable with the process.
3. Cut a base page larger than normal with the Passport page boundaries marked in chalk. This gives you more freedom to move. Add a layer or two of stabilizer. Take two to five minutes and doodle some images on the page. There are no right or wrong subjects or shapes. When you are finished doodling, trim the page to 5" x 7".

landscape sketching

Small or large, the process is the same. Creating a fabric base for a landscape sketch lets the background fabrics do some of the color work.

1. Draw a simple landscape sketch on your fabric. Prepare chosen fabric scraps by applying fusible web to the back. Cut out the desired patterns (trees, flowers, rocks) and fuse them to your base fabric along the sketch lines.
2. Add stabilizer to the back of the fabric and start stitching. Work from the background to the foreground. Change thread colors as needed. Don't be afraid to splash colors in unexpected places.
3. Don't forget a signature. Attach the finished landscape to your passport page.

thread & media

Explore the mixed media world by adding color to your thread sketches.

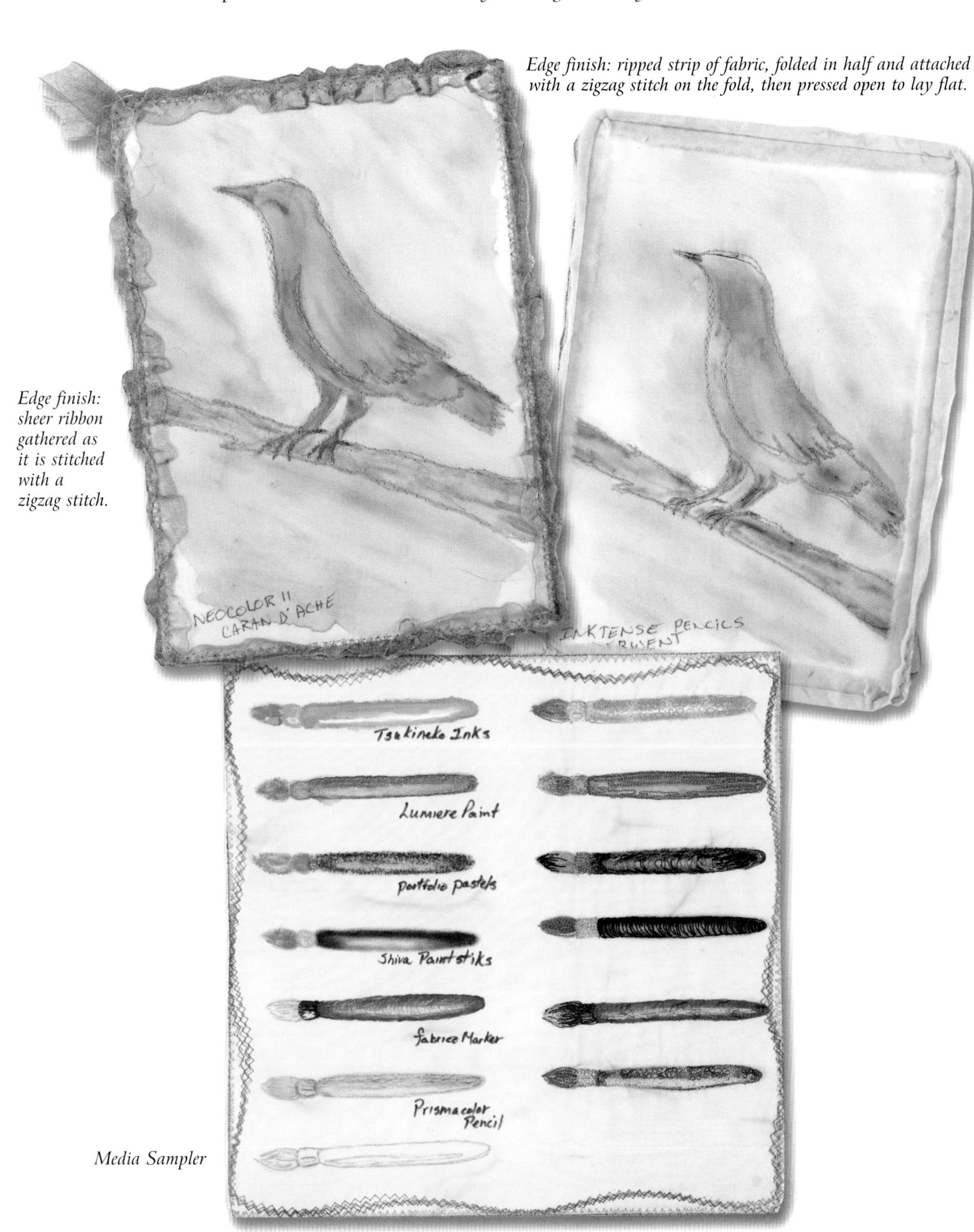

Edge finish: ripped strip of fabric, folded in half and attached with a zigzag stitch on the fold, then pressed open to lay flat.

Edge finish: sheer ribbon gathered as it is stitched with a zigzag stitch.

Media Sampler

getting started

Some colorants to use on your fabric include colored pencils, watercolor pencils, oil paint sticks, water-soluble wax pastels, fabric or acrylic paints, acrylic or fabric inks in liquid, pencil or marker form, transfer paints and dyes, markers, and pens. Not all products are permanent on fabric so testing is a great idea. Try them out on one of your thread sketches or make a media sampler, as shown on page 84.

1 Lightly brush paint on the background page. After the paint dries, press the fabric.

2 Free motion stitch over the paint, letting the contours of the paint determine the design.

Try This!

Edge finish: zigzag stitch in a metallic thread.

1. Using a heavy weight thread, fill a page with moderately spaced grid or meander stitching.
2. Lightly brush the surface of the fabric with one or more colors of fabric paint, allowing the paint to catch the raised areas between stitches. The heavy thread line will depress the fabric and allow the non-stitched areas to stand out.
3. Add a decorative pin to finish the page.

Trace the bird pattern and transfer it to multiple page bases. Experiment with various colorants to determine what works best for you. Label each page with the type of colorant used. Refer to the Passport pages on page 84.

thread painted appliqué

Thread painted appliqués are created separately and then appliquéd in place. This eliminates distortion on the background fabric.

Edge finish: a zigzag stitch and a decorative wavy line in heavy metallic thread.

getting started

Appliqués can be created on any base fabric including cotton, tulle, and organza or it can be made strictly of thread stitched on water-soluble stabilizer, as shown in Thread Lace Grid on page 110.

Machine set up:
Free motion
drop feed dogs

Tension:
Upper loosen slightly

Needle:
Embroidery 90/14

Top thread:
Assorted medium weight threads

Bobbin:
Fine

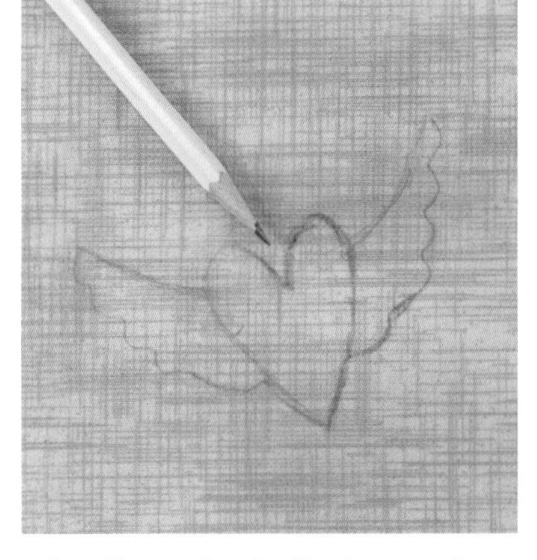

1 Draw a heart with wings or other image onto a piece of fabric.

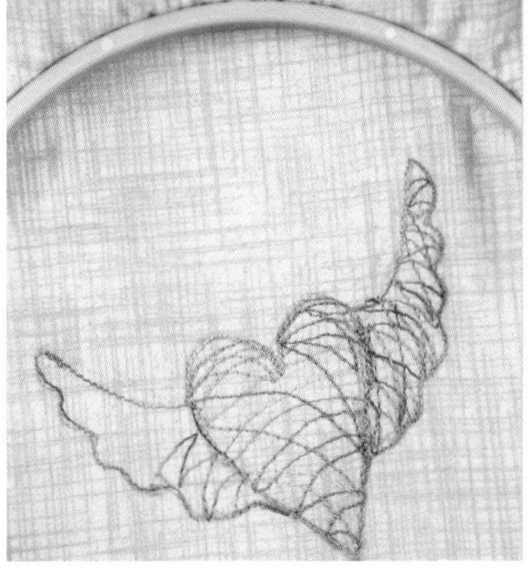

2 Layer the fabric over a non-fusible stabilizer and place in an embroidery hoop.

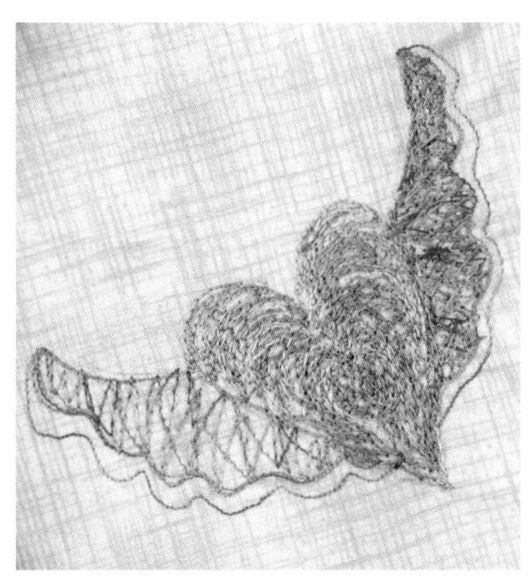

3 Referring to Thread Painting on page 66, outline stitch the shape and then lightly stitch the interior. Continue stitching layers of thread working back and forth between areas and colors. When finished stitching, remove the fabric from the hoop and press from the back. Cut out the appliqué shape and stitch in place on your fabric base page.

4 To make a hanging accent for your page create an appliqué following steps 1 - 3. Layer the appliqué over a matching piece of fabric or felt. With the two pieces wrong sides together, lay a piece of ultra heavy thread or fine cord around the perimeter. Use a matching cord and hand couch in place to cover the raw edge between the two fabrics. Refer to Couching on page 90.

5 Finish with a cord or ultra heavy thread loop and attach to the page.

Just for fun

A bird image was rubber stamped on tissue paper and outline stitched before tearing the tissue away and continuing with the Thread Painted Appliqué steps.

the techniques

couching

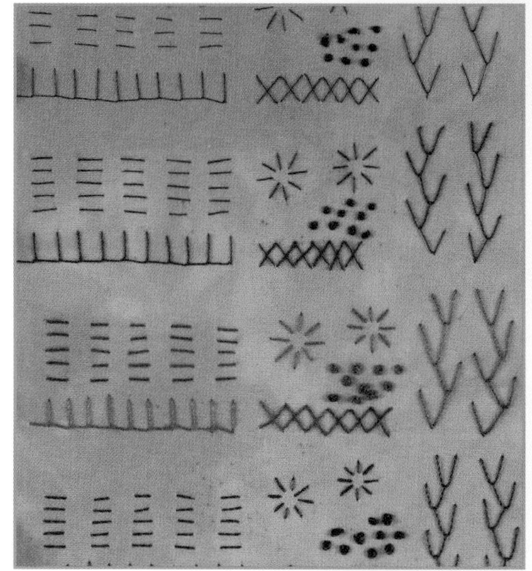
hand stitching

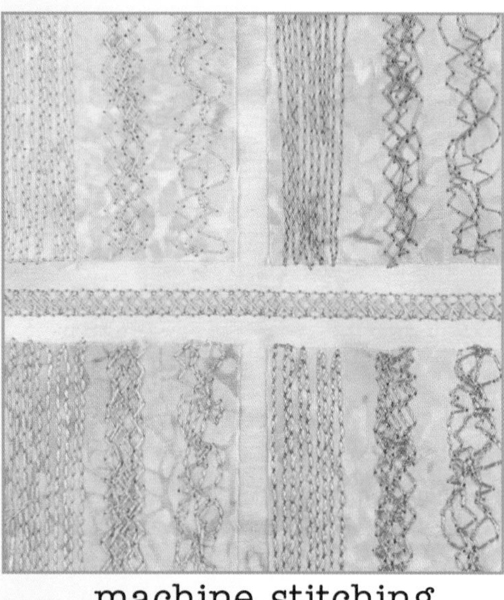
machine stitching

confetti fabric

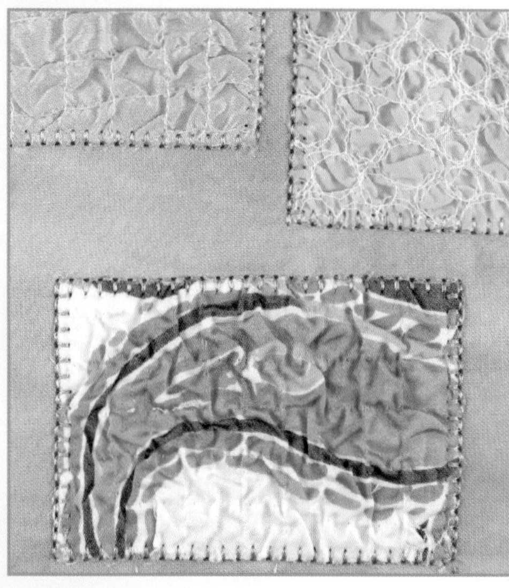
heat shrinking thread

faux weaving

Thread as Texture

When we start thinking of thread as a way to add texture and layers of interest to our fabrics an exciting world of possibilities opens up. Couching, hand embroidery, and funky free motion stitching all add layers of color and texture on top of fabrics and paper. Programmed stitches and heat shrinking thread let us build intriguing textured fabrics from stray scraps and thread bits. You will never look at your scraps the same way again!

couching

Couching enwraps fiber with stitches landing on either side of the fiber rather than passing through it. Couching stitches can be a simple zigzag or blanket stitch, a decorative or utility stitch, a plain hand stitch or one that incorporates beading.

Edge finish: light zigzag stitch and couched sari waste silk fiber.

Edge finish: two strands of heavy thread couched with a zigzag stitch

getting started

Use a variety of cords, thread and yarn, pre-strung beads, craft wire, or natural materials for this Passport page. We used a zigzag stitch as our couching stitch.

Machine set up:
Normal

Tension:
Balanced

Needle:
Universal or sharp sized to top thread

Top thread:
Multi-color cotton

Bobbin:
Medium weight

Foot Notes:
Couching foot or satin stitch foot

Refer to your sewing machine manual for information on your machine's specific specialty sewing feet. See page 13 for examples of couching feet.

1 Cut two heart shapes out of two different fabrics, one slightly larger than the other. Apply fusible web to the back of both hearts. Fuse them to the fabric page base with the smaller heart on top of the larger one. Using a marking pen or chalk, draw two 'wonky' shaped rectangles to frame the hearts. Leave extra space at the bottom to add beads and wire. Lay an ultra heavy thread or other fiber in place along the smaller heart's perimeter, leaving a small tail. Pin in place if necessary.

2 Adjust your stitch width to fall just to either side of the thread or fiber you are couching. Pivot around the heart and cross the beginning and ending tails at the point of the heart. Trim the tails after taking a locking stitch. Couch another fiber around the larger heart. Start at the lower point and leave a 3" tail. When you have reached the starting point, stitch around the thread tail in a curlicue shape, pivoting frequently. Take a locking stitch and snip the threads. Start at the point of the heart and stitch around the original 'starter' tail making a mirror image of the first curlicue shape.

Create a reference page

Edge finish: blanket stitch.

1. Mark chalk lines across the page approximately 1/2" to 3/4" apart. These will be guidelines to couch the fiber samples on the page. The upper portion of the page shows ultra heavy thread, fine yarn, bulky yarn, hand-dyed rayon satin cords, plastic canvas lacing, leather lacing, and machine wrapped cords. Refer to Machine Wrapped Cords & Sticks on page 138. Stitching over satin cord is on the lower portion.
2. Select your fiber and couching thread. Take a few locking stitches in the center if needed to secure the fiber. Skip this step if couching leather lacing. Set the zigzag width to snugly couch the fiber and stitch to the end. Secure if necessary. The first three samples were stitched in multi-color cotton. The others were stitched with invisible thread.
3. Repeat step 2 for the stitch samples using stitches available on your machine. We used zigzag, blanket, and decorative stitches, as well as hand stitching.
4. Make labels for your page by rubber stamping letters on fabric. Refer to Rubber Stamping on page 170.

3 Choose a fiber to couch along the first and second frames. Set stitches to the appropriate length and width to wrap the fiber snugly. Leaving a short tail at the corner, couch the fiber along the frame outline. Take a locking stitch when you return to the starting point to hold the fiber in place. Trim both tails flush with the frame.

The inside frame on our Passport page is a satin cord couched with a contrasting thread. The outer frame is a machine wrapped cord stitched with a blending thread. Refer to page 138, Machine Wrapped Cords & Sticks.

tips

- Use a thread that will be strong enough to hold the fiber in place.
- Stitches should be wide enough to clear the fiber, but be snug to it. The longer the stitch length, the less thread will show. Shorter stitch lengths can give a burst of thread color for special effect.
- Decorative stitches can be used. You may wish to choose a stitch that doesn't overly pierce the fiber.
- If the beginning or ending of the couching will not be secured in a seam, pierce the fiber with a few stitches to hold it in place. Or, in cases where the back will be covered, use a large eye needle to pull the fiber to the back of the work and secure with a small piece of fabric fused over the fiber.
- When couching around a curve, pivot every 1 to 3 stitches to keep a smooth line. When going around a corner, take a few extra stitches at the inside corner and then pivot the fiber and needle 90-degrees.
- When couching around a corner, pivot with the needle down on the inside of the corner to create a smooth turn. With some fibers you may need to use a pin to pull a tiny bit of the fiber behind the foot so it doesn't pull too sharply around the turn.
- Hand stitching can be a simple whip or couching stitch. It is also a great opportunity to add beading while stitching.

Try this

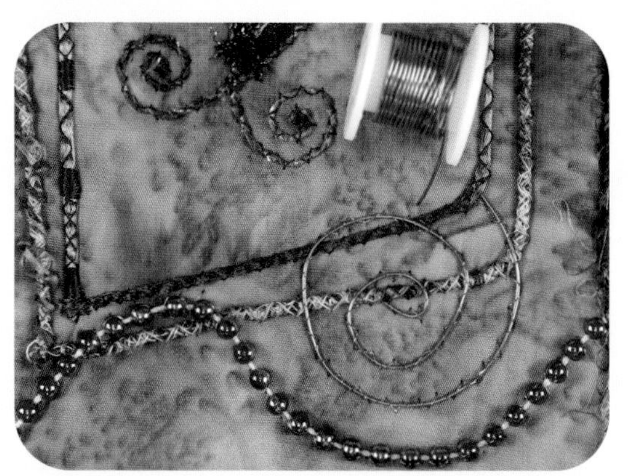

- Create interesting bands of stitching with a contrasting thread color and varying the stitch length. Start with a short (.35) satin-like stitch, then lengthen the stitch to about 3.0 for a few stitches and reduce again to .35. Repeat to create a banded effect along one side of the frame.
- Couching a stitch wrapped cord in the same thread it was created with makes the couching stitches disappear.
- Add beads as you hand couch a cord. Take three stitches without beads, and then repeat the beaded stitch.
- Use a specialty bead couching foot to couch pre-strung beads with clear monofilament thread.
- Pre-shape craft wire and couch it to your page with metallic thread.

Journal

Embellish a purchased journal with silk paper, whip stitching, hand stitching, free motion stitching, and beading. Ripped fabric strips are couched on the spine.

Hand Stitching *Couching* *Free Motion Stitching*

hand stitching

Hand stitching and embroidery allow us to slow down and get closer to our work. We love the relaxing and meditative aspects, too.

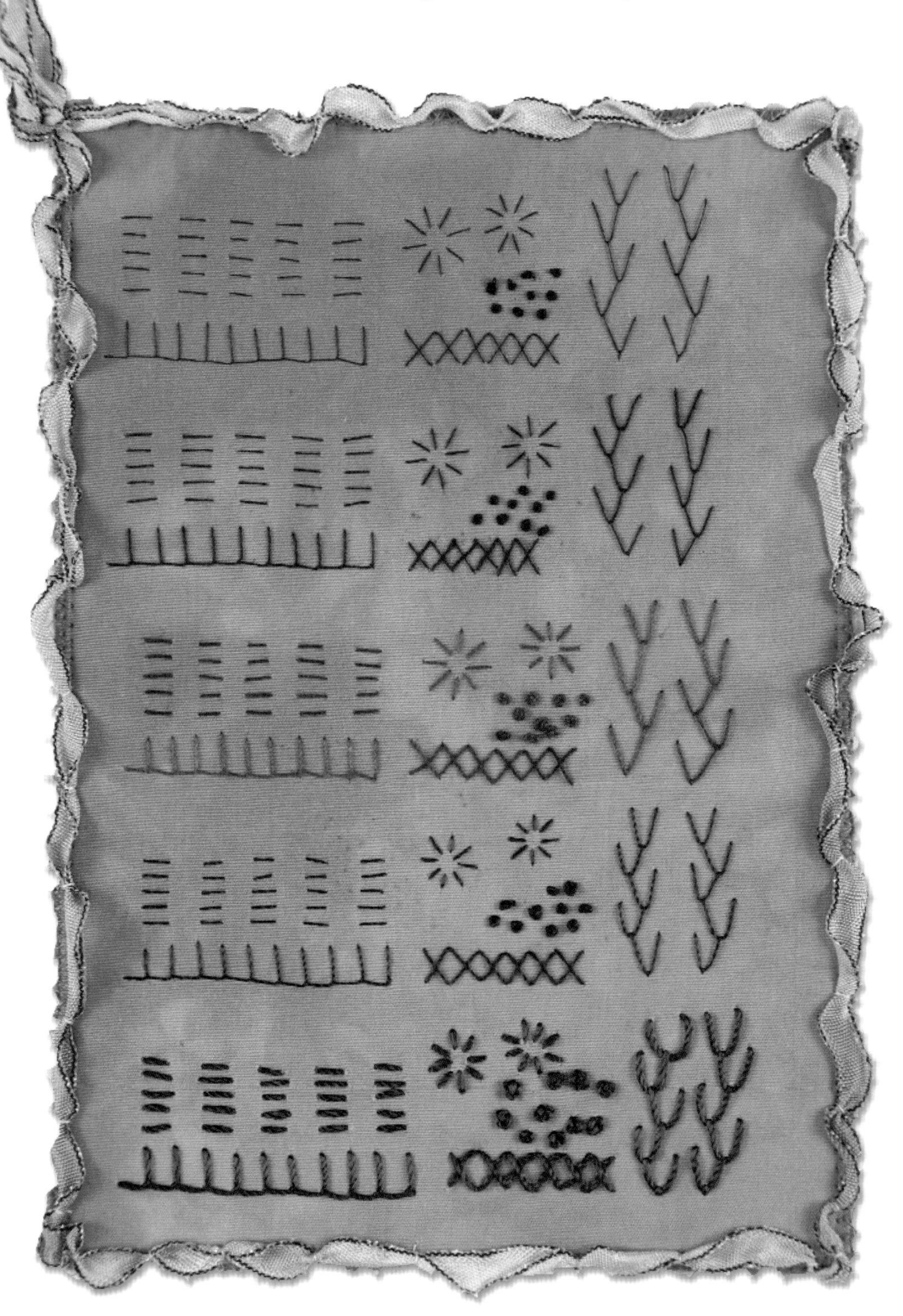

Edge finish: zigzag stitch with ribbon hand couched over it.

getting started

Divide your Passport base page into five sections with marking chalk or a fabric pencil. Choose five embroidery stitches to use on your sampler. Ours uses a running stitch two ways - straight and arranged in a star pattern - buttonhole stitch, cross stitch, French knots, and the feather stitch. Stitch each pattern in the different weights and fibers of thread.

Threads:
Five threads in varying weights, fibers, and colors

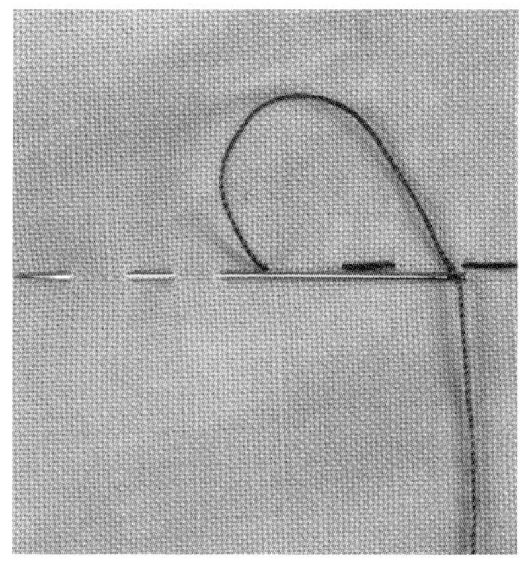

The Running Stitch

is worked from the front of the fabric and from right to left. Take the needle in and out of the fabric at relatively equal distances. You can load two or three stitches at one time.

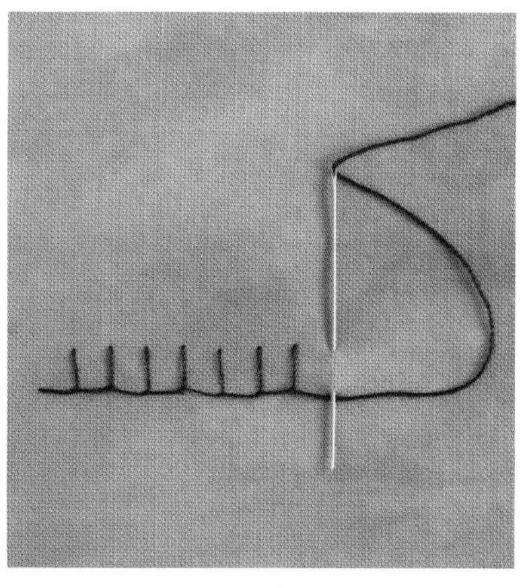

The Buttonhole Stitch

(also referred as a blanket stitch) is worked from the top of the fabric from left to right. The stitch resembles a backward letter L. Bring the needle up along the bottom edge where you plan to stitch. Put the needle into the fabric about 1/4" away on a diagonal. Bring the needle up on the bottom edge looping the thread under the needle.

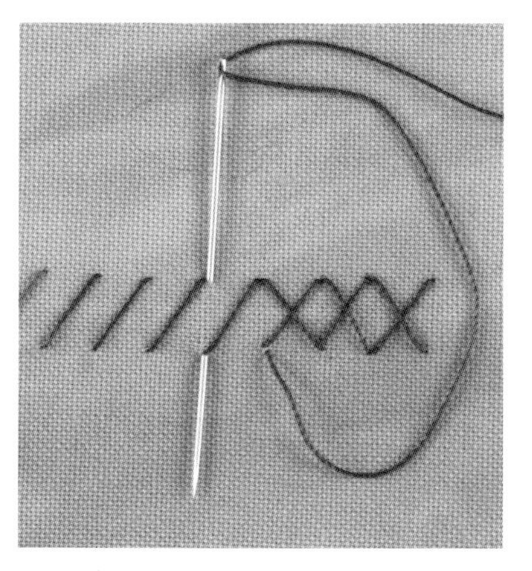

The Cross Stitch

may be worked individually or in a row. Half the stitch is worked from left to right and the 'cross' stitch is worked from right to left on top of the row. Bring the needle up at the lower left edge of where you want the stitch to begin. Put the needle back in the fabric about 1/4" away on the diagonal. Bring the needle back up directly below where you went in and on the same horizontal line. Continue until the row is the desired length. Then reverse directions and cross the previous diagonal line going in the opposite direction.

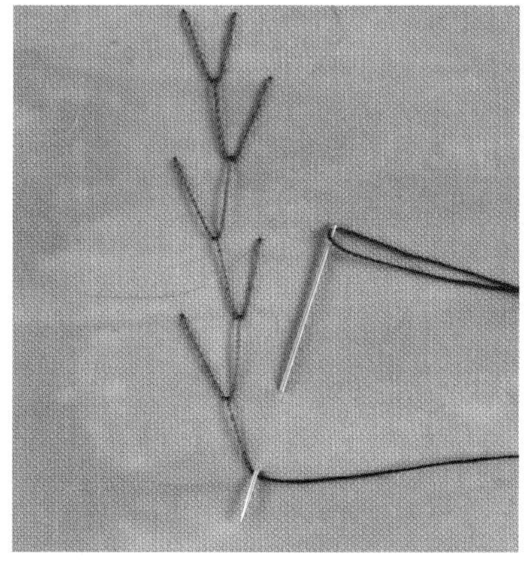

The Feather Stitch

is worked on top of the fabric moving from top to bottom. Bring the needle up from the back at the upper left corner of where you want your stitch to begin. Put the needle back in the fabric approximately 1/4" away and bring it back out approximately 1/4" below and in the middle of the first two stitches, catching the thread under the needle and forming a 'V' shape. From this point put the needle in the fabric 1/4" away on either the right or left side of your first stitches and repeat the stitch until the desired length is achieved.

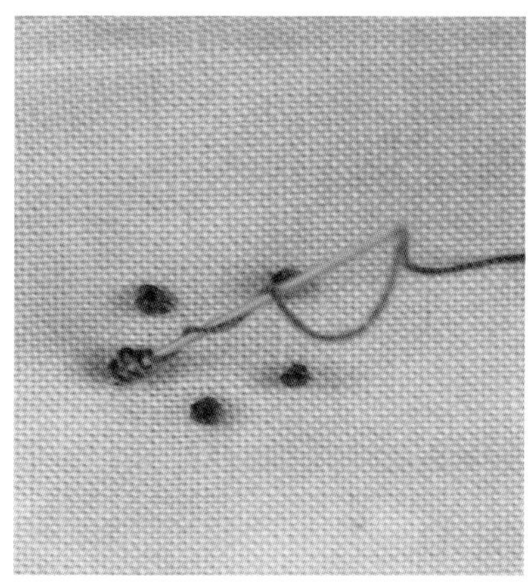

French Knots

are worked one at a time. Bring the needle up from the back of the fabric, wrap the thread around the needle two to five times. Hold the thread snugly in your left hand while you put the needle back into or immediately adjacent to (one thread away) the hole you came out of. The more times you wrap the thread around the needle the larger the knot.

Playing with stitches

Sometimes you just have to break the rules. Play with the size, shape, density, and pattern of embroidery stitches. Each traditional stitch offers almost limitless variations with which to experiment and play. Choose three or four different stitches. Our sample uses running stitch, feather stitch, blanket or buttonhole stitch, and cross stitches.

1 Running stitch can be used to create different patterns.

2 Ricing, also called seed or scatter stitch, is a fun stitch to use to fill space and add texture.

Feather stitch play yields fun texture as well as forms shapes that reflect nature, like branches, vines, underwater plants, and netting.
3 Wonky sides
4 Short and fat
5 Unbalanced
6 Layered with a row of metallic added

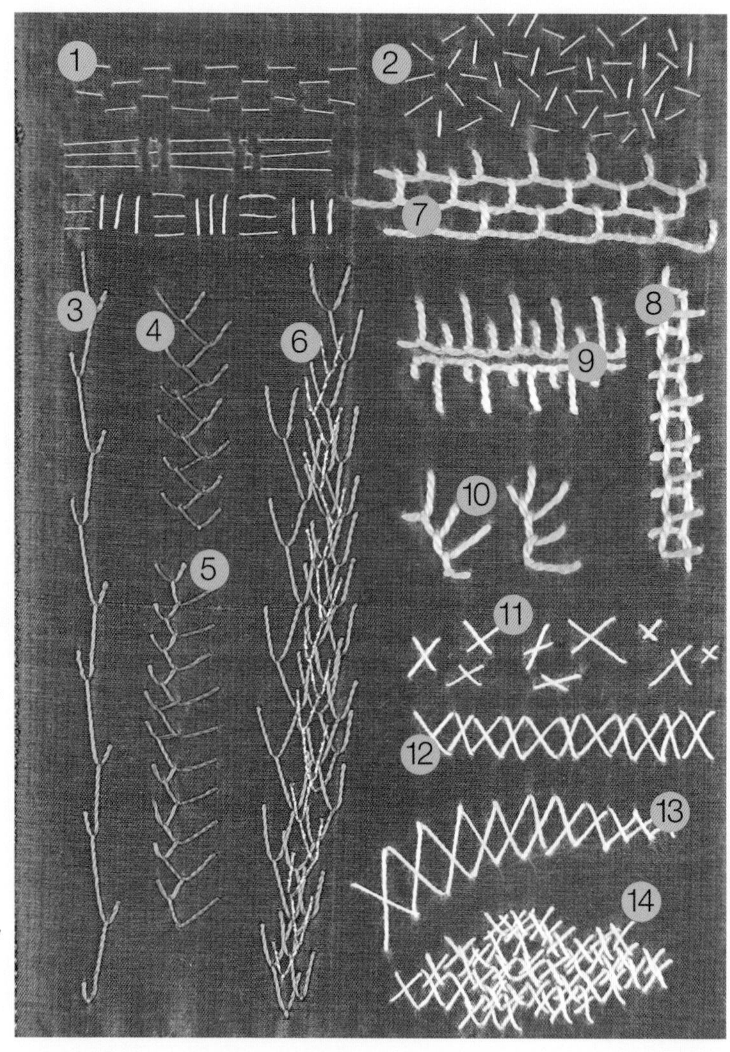

Buttonhole stitches can lend themselves to both geometric shapes and organic, natural shapes.
7 Bricked
8 Mirror image and overlapped
9 Mirror image and varying height.
10 Curved and varying height

Cross stitches are great for adding motion and texture.
11 Variations of side lengths
12 Alternating wide and narrow stitches
13 Undulating stitches
14 Various sizes layered

tips

- Iron a light weight fusible interfacing to the back of your fabric to give it more body when hand stitching. This is especially helpful when using silk and other light weight fabrics.
- Hand embroidery hoops are a personal choice. Try stitching with and without one to discover which you prefer.
- If you prefer a hoop do not cut your fabric to Passport page size until after you finish hand stitching. Mark the finished size of the page with a fabric pencil, chalk, or marker.
- Variegated threads are wonderful to use with hand stitching.
- A temporary strip of masking tape can help keep your stitching line straight.

machine stitching

Enrich the surface texture of your fabric with simple stitching. Different properties of thread, how fast you move the fabric and the direction you move the fabric allow for an incredible array of variations. Try different thread weights and fibers as well as blending and contrasting colors. When choosing a programmed stitch to use with free motion stitching make sure the swing of the needle is not larger than the width of your free motion foot.

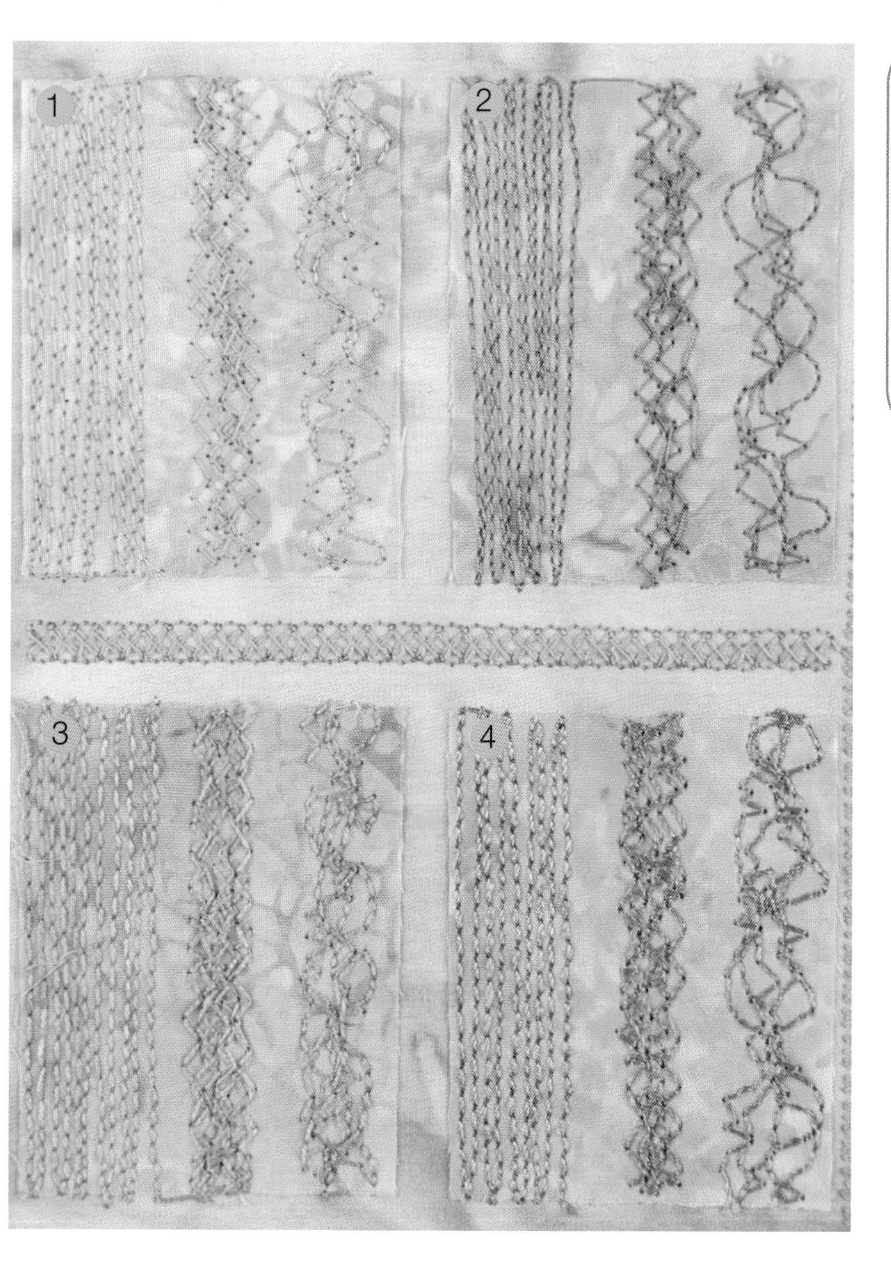

Machine set up:
Free motion
drop feed dogs

Tension:
Balanced or slightly loosen upper

Needle:
Embroidery 90/14

Top thread:
Any medium and
heavy weight decorative

Bobbin:
Ultra fine and fine polyester

Cut four 2" x 2-1/2" rectangles of fabric. Fuse to a fabric page base. Back the fabric with a stabilizer. Each rectangle is stitched with a free motion straight stitch, free motion zigzag stitch, and free motion programmed stitch. Refer to the Stitch Dictionary on page 22.

1 Use a medium weight thread that blends with the fabric to stitch the first rectangle.

2 The second rectangle is stitched with a contrasting medium weight thread.

3 The third rectangle is stitched with a blending heavy weight variegated thread.

4 The fourth rectangle is stitched with a contrasting heavy weight thread.

confetti fabric

Confetti fabric looks like a complex surface but is simple to create. Use it for appliqué shapes or as a focus fabric for small projects such as journal covers, jewelry, and whimsical creations.

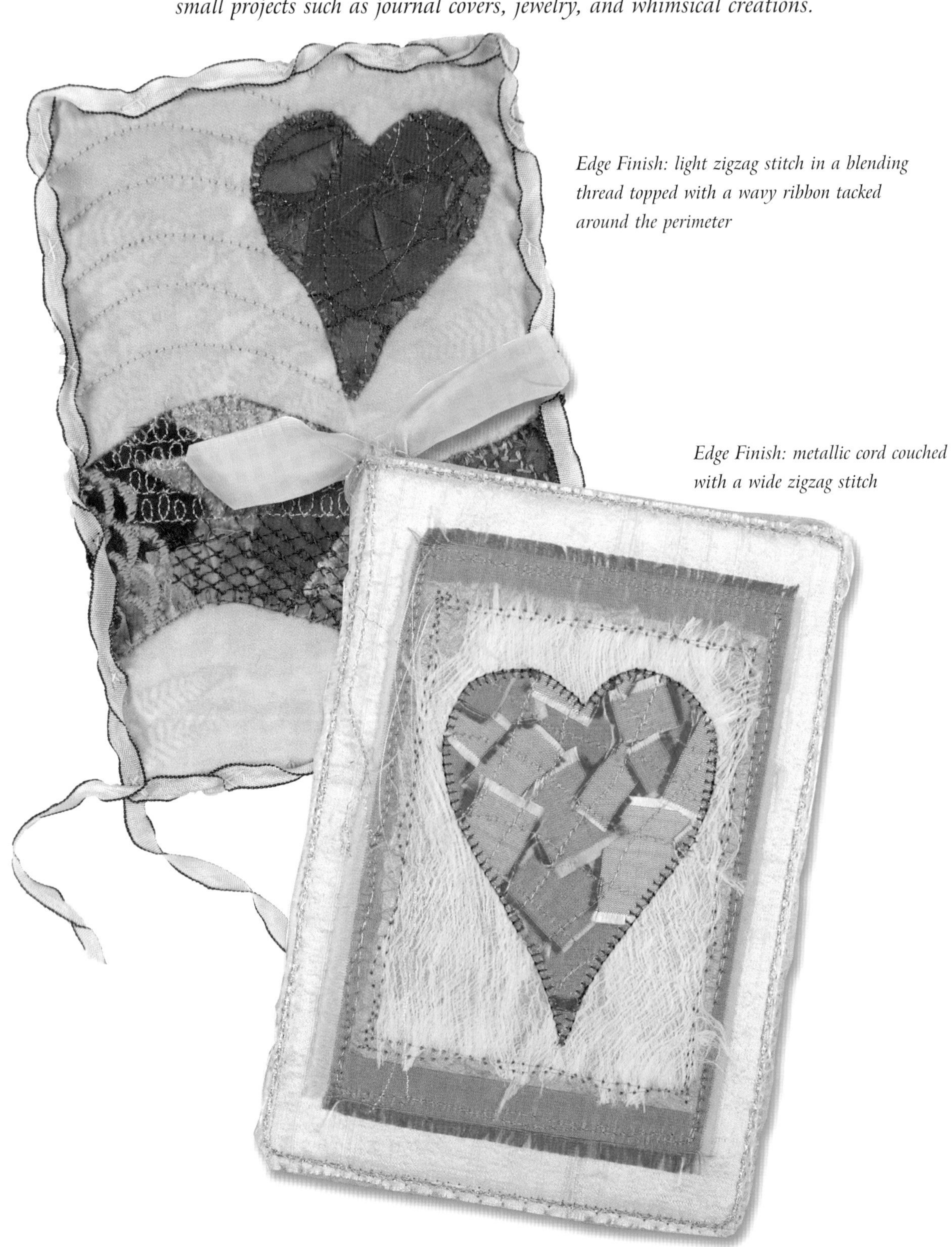

Edge Finish: light zigzag stitch in a blending thread topped with a wavy ribbon tacked around the perimeter

Edge Finish: metallic cord couched with a wide zigzag stitch

getting started

Shot silks are a good choice for SINGLE FABRIC CONFETTI, as the frayed edges highlight the different colors of the warp and weft weave. Work at the ironing station to avoid fluttering fabrics. Work on a base fabric of light weight, fusible stabilizer or fusible backed fabric to bind the fabric bits.

Machine set up:
Normal

Tension:
Balanced

Needle:
Embroidery 90/14 or Metallic

Top thread:
Fine or medium weight decorative

Bobbin:
Fine or medium weight

1 If necessary, layer your backing fabric with fusible web. Fuse it in place following the manufacturer's directions and remove the paper backing. Lay fabric or stabilizer fusible side up. Create confetti by scattering fabric bits over the exposed fusible web. Cover the base with one layer of slightly overlapping fabric pieces.

2 Fuse the fabrics to the stabilizer by covering the confetti with parchment or a pressing sheet. Be aware of the type of fabric being fused when setting the heat on the iron. If any of the bits lift or you see bare spots, drop a confetti bit over the spot and re-fuse.

3 Choose normal or free motion sewing to stitch wavy lines across the confetti fabric. Use a straight or decorative stitch. Try to catch any floppy raw edges that remain unfused. Stitch with one thread, then criss-cross the lines with another. Don't over-do the stitching.

4 Place the heart template on page 100 over the confetti fabric until you find a placement you like. Mark the shape with chalk and cut it out.

5 Place the shape on the background fabric and stitch in place with a blanket stitch in a blending color.

Mending your heart

If you discover a bare spot where stabilizer peeks through, fuse or stitch another piece of fabric over the spot. You may also stitch over the spot using a blending color, but it sounds like an opportunity to try something new.

confetti fabric

Three to five fabrics work great for MULTI-FABRIC CONFETTI. The process is similar to single fabric confetti, but with a greater number of fabrics, threads, and stitches.

Machine set up:
Normal

Tension:
Balanced

Needle:
Embroidery 90/14

Top thread:
Any decorative thread

Bobbin:
Fine or medium weight

1 Apply fusible web to one side of stabilizer or base fabric. Place fusible side up on the ironing board and cover with fabric bits in multiple colors. Try not to let too many of the same fabric bits touch each other. When the entire stabilizer is covered and the fabric placement is to your liking, fuse in place.

2 A scrap of fabric is handy for trying out stitch and thread combinations. Better to stitch them out than rip them out. Stitch each pattern and thread combination in a grid or curving lines until all of the confetti is stitched over.

3 Position the wave template below on the fabric, mark the outline, and cut out.

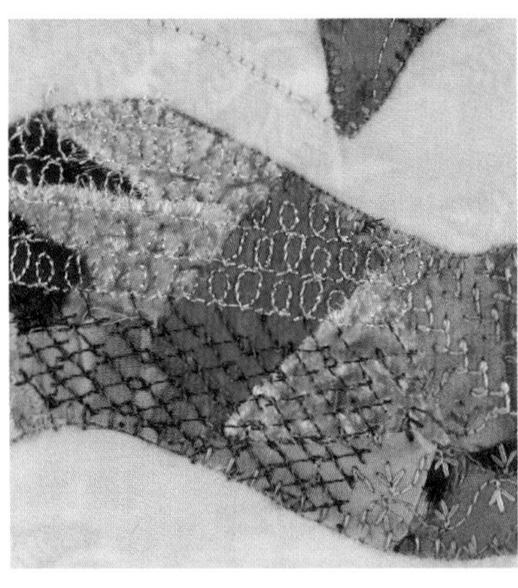

4 Stitch the wave appliqué in place below the heart using a blanket stitch and a blending thread.

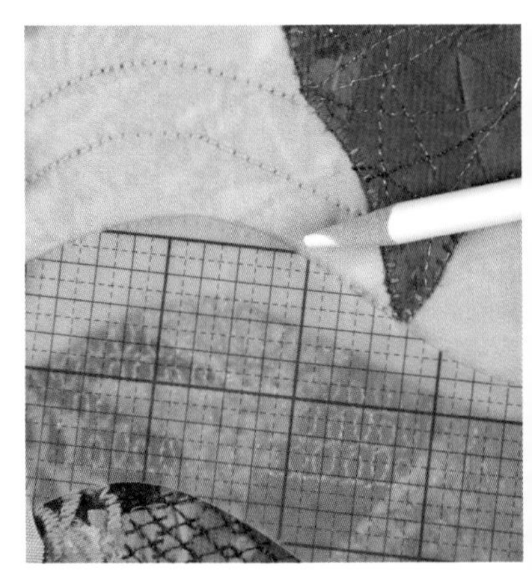

5 If you like, you can repeat the wave motif by marking all or part of the stencil onto the open background. The wavy lines on this page are stitched in a soft wool thread.

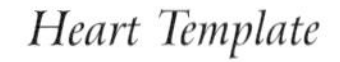

Wave Template

Abstract Square

A confetti fabric base is covered with painted heat distressed stabilizer to add interest and texture. The irregularly stitched red solid piece is mounted on a black background and stitched in a regular grid.

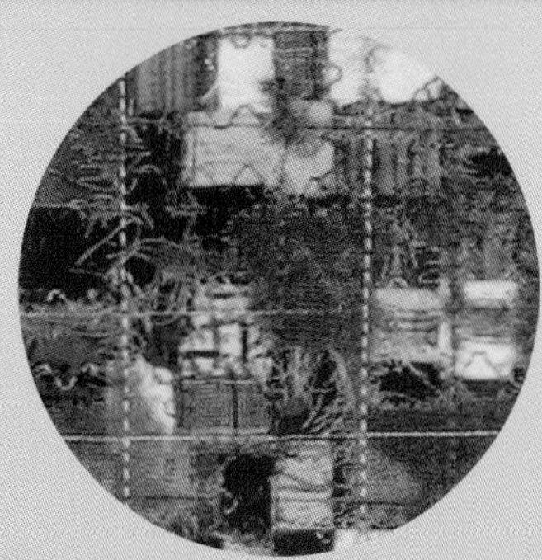

Confetti Fabric;
Painted Heat Sensitive Stabilizer

heat shrinking thread

Heat shrinking thread allows you to create interesting textured fabrics easily. These fabrics can be used on their own and are wonderful textural embellishments on garments, mixed media collages, and home décor.

Edge finish: zigzag over ribbon knotted at intervals.

getting started

Machine set up:
Free motion
drop feed dogs

Tension:
Balanced or slightly loosen upper

Needle:
Embroidery 90/14

Top thread:
Heat shrinking or light weight (this will result in less shrinking of the fabric)

Bobbin:
Heat shrinking

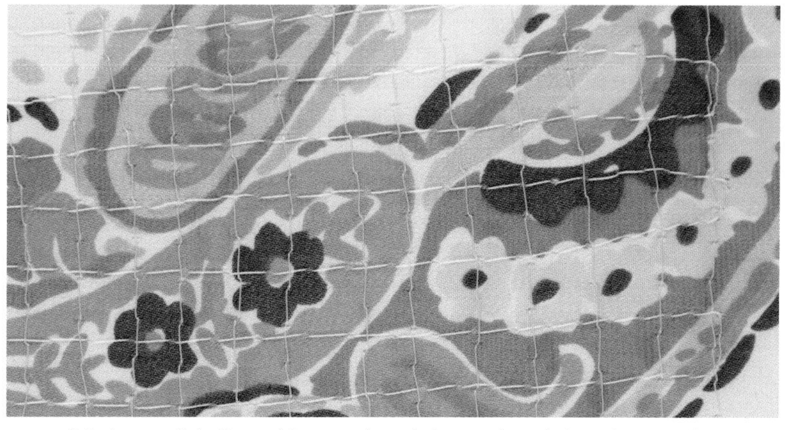

1 Stitch a straight line grid, meander stitch, granite stitch, or in wavy lines. Use a machine embroidery hoop to help hold the fabric taut.

2 Remove the fabric from the hoop. Set your iron to a cotton steam setting. Hold the iron just above the surface of the stitching, moving across the surface without touching the fabric. The stitching will begin to shrink almost immediately. If not, use more steam. Steam both the front and the back of the fabric.

3 To secure the textured surface, fuse a piece of light weight interfacing to the back of the fabric. Leave the stitching in or take it out as desired.

heat shrinking thread

Try different weights of fabric in sheers, silk, and cotton to create a textured fabric out of scraps.

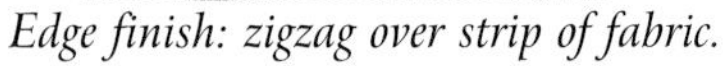

Edge finish: zigzag over strip of fabric.

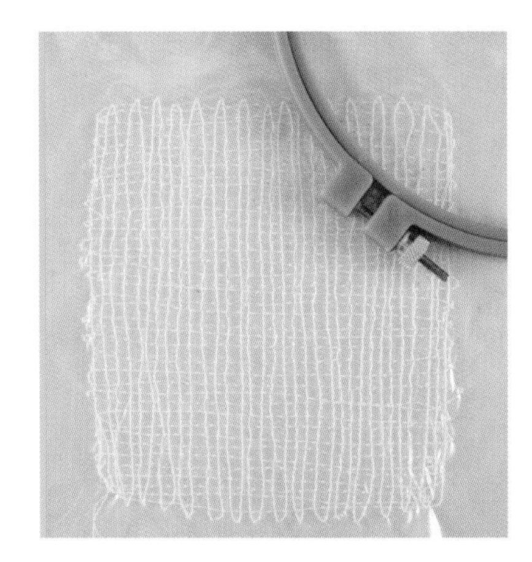

1 Secure a piece of heavy duty water soluble stabilizer in a hoop and stitch a grid. We stitched a very tight grid with stitch lines about 1/8" apart for maximum shrinkage. Gather a pile of fabric scraps (cotton, silk, linen, or polyester), ribbons, yarns, and threads.

2 Lay the fabric scraps directly on top of the stitched grid.

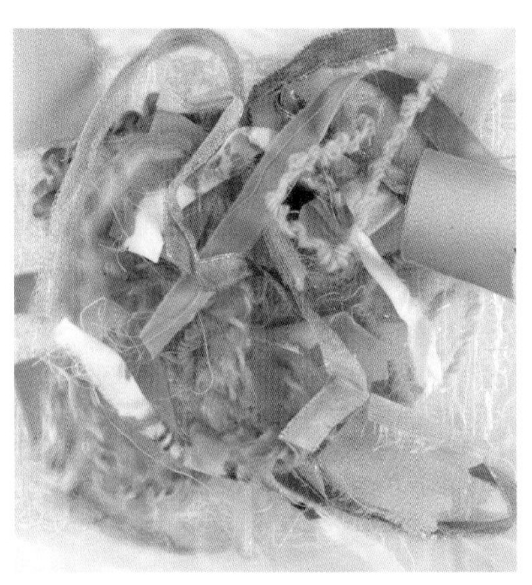

3 Add fibers, yarns, threads, and ripped strips of fabric to cover most of the grid. Use an assortment of textures.

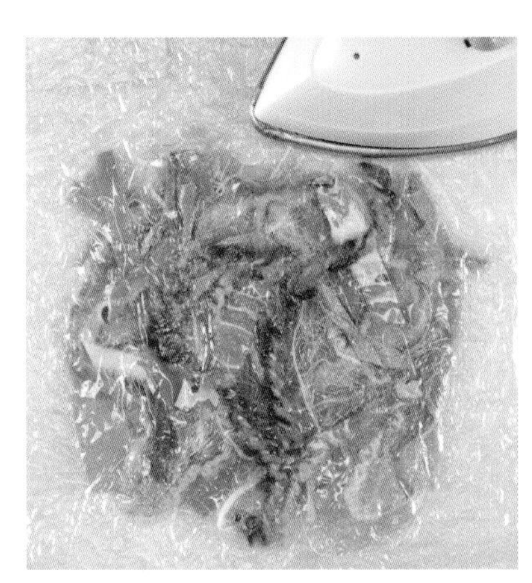

4 Lay a second piece of water soluble stabilizer on top to sandwich the fabric and fibers in the middle. Use a light or medium weight water soluble stabilizer. Set your iron on medium heat and lightly touch it to the layers of stabilizer around the edges. This will seal the edges so the fibers do not shift too much.

5 Stitch over the encased fibers. We used a light weight thread in both the top and bobbin and stitched wavy lines in a loose grid. A meander stitch would also work well.

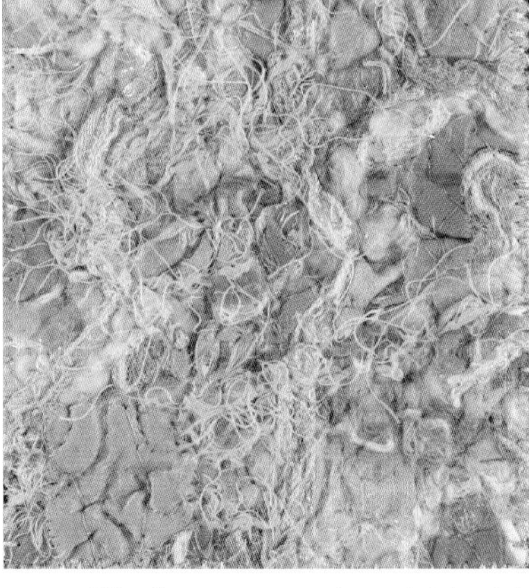

6 Trim the excess water-soluble stabilizer and rinse out thoroughly. Let the piece dry fairly well. Turn upside down on an ironing surface. Set the iron for steam and use a medium high to high setting. Hold the iron just above the surface of the heat shrinking thread. The thread will shrink and you will be left with a wonderful textured surface. Fuse a piece of light weight interfacing to the back and trim as desired.

Color Burst

Color effects can be layered in threads. Similar to stitching the color wheel, work in one light layer and continue building up the layers to create a beautiful burst of color.

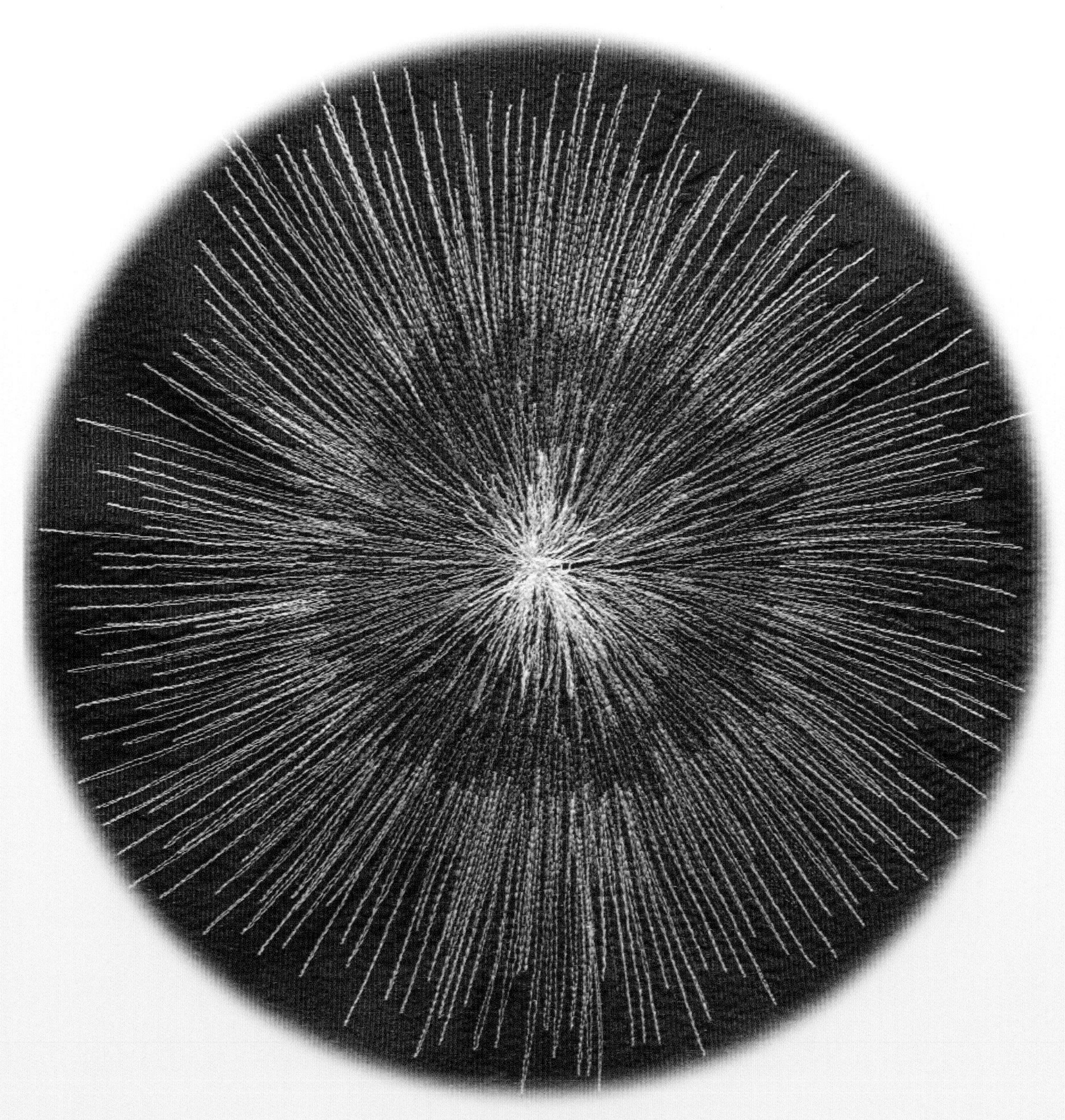

Thread Painting

faux weaving

Faux weaving is a quick and easy technique you can color coordinate with any project.

Edge finish: zigzag stitch in metallic thread

getting started

Make the weaving card and window any size you want. Weave any threads or fibers you choose and stitch with any type of thread.

Machine set up:
Free motion
drop feed dogs

Tension:
Balanced

Needle:
Embroidery 90/14

Top thread:
Heavy thread and decorative thread

Bobbin:
Medium weight cotton or polyester
in complementary colors to top

1 Cut a 3" square out of heavy cardboard. Cut a 1-1/2" square window from the center of the 3" cardboard square. Tape the starting end of the wrapping thread in place. Begin winding the thread around the cardboard square from top to bottom. The up/down winding of the thread is the warp thread. Wrap with a very light hand to minimize distortion. Tape the ending tail in place. Repeat this for the side-to-side, or weft of the weaving. Do not worry if the card bends.

2 Free motion stitch all the threads that overlap the window opening. Be careful not to stitch into the cardboard. As the weaving threads are stitched, they will gather in small bunches. Snip the outside edges away from the card.

3 Create your Passport page by layering fabrics or paper under the faux weaving and stitch with decorative thread if desired.

the techniques

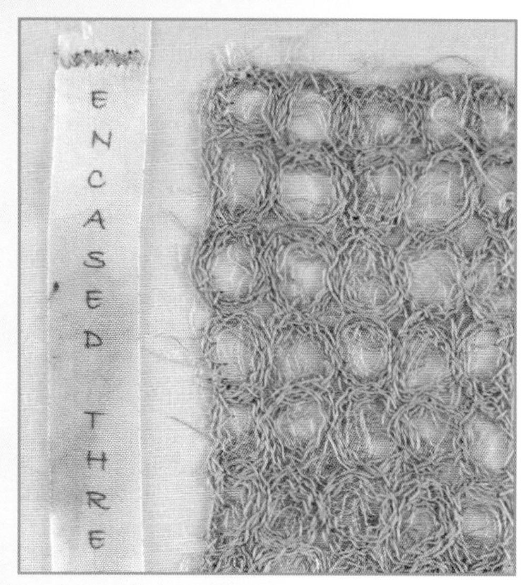

thread lace grid

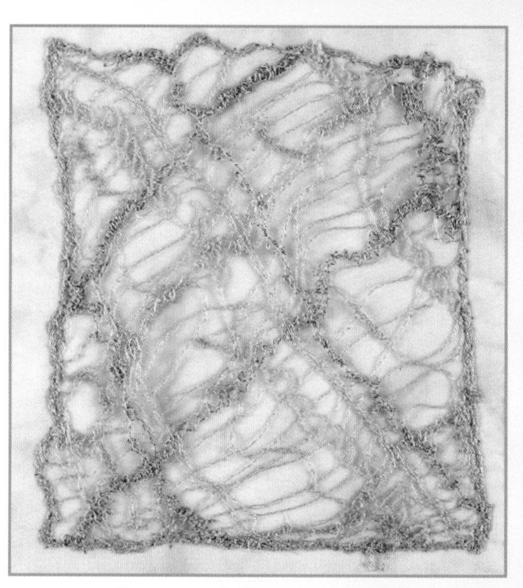

thread lace grid

thread lace elements

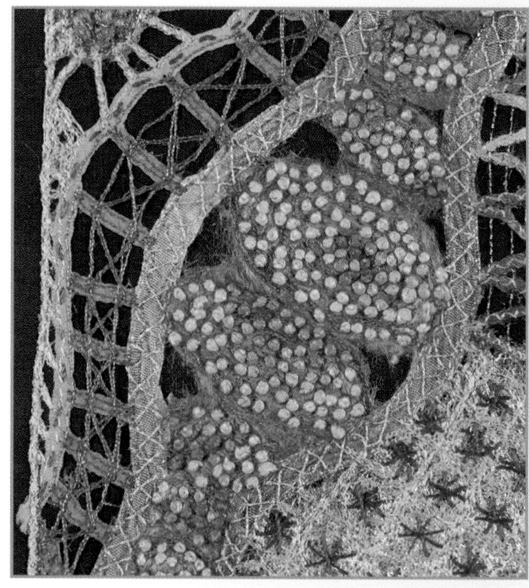

free form embroidery

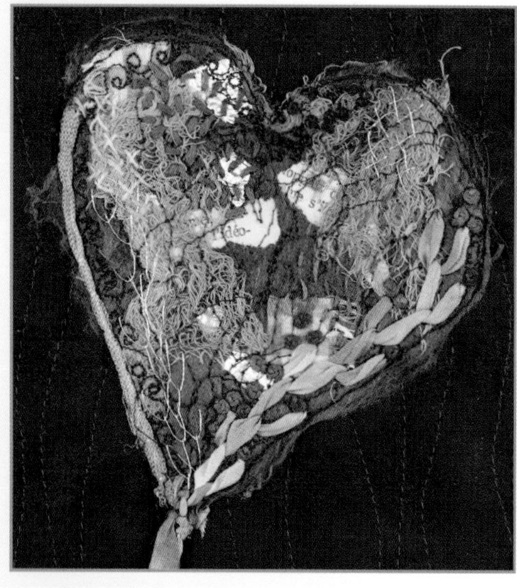

free form embroidery

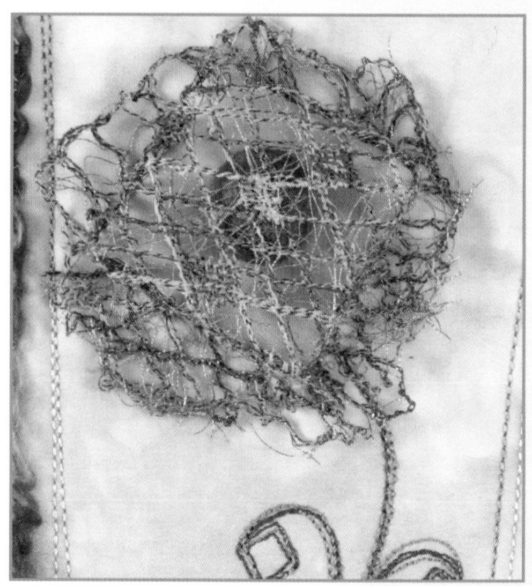

small thread structure

Thread as Structure

Thread is the true star of the show when creating thread lace, appliqués, and structures such as vessels and bowls. Water-soluble stabilizers allow us to create unique fabrics and laces, without weaving, knitting, or tatting. The variety of intriguing effects you can achieve with these stabilizers is amazing. Imagine being able to add beading and hand embroidery to delicate wisps of fabric, fibers, and thread. We find the process magical and know you will too.

thread lace grid

To create a lace fabric out of thread you must work in a grid system. There are so many possibilities with this technique it just may become one of your all-time favorites.

Edge finish: ripped strip of silk couched with a zigzag stitch

Edge finish: torn strip of fabric sewn on with a straight stitch

getting started

The threads you stitch for the BASIC THREAD LACE GRID must interconnect with each other at multiple points to create a strong fabric that won't fall apart once you dissolve the supporting stabilizer. Your underlying grids need not be evenly spaced linear lines. They can be circular, triangular, or organic shapes, evenly or unevenly spaced apart. Be sure to give the stitches something to hold onto after the stabilizer is dissolved.

Machine set up:
Free motion
drop feed dogs

Needle:
Embroidery 90/14

Tension:
Loosen top

Top thread:
Decorative cotton, polyester, silk, or rayon

Bobbin:
Light weight or matching

1 Cut a piece of water-soluble stabilizer slightly larger than your embroidery hoop. Draw a pattern or grid on a piece of paper. Lay the stabilizer on top of the design and trace the design on the stabilizer with a pen or marker.

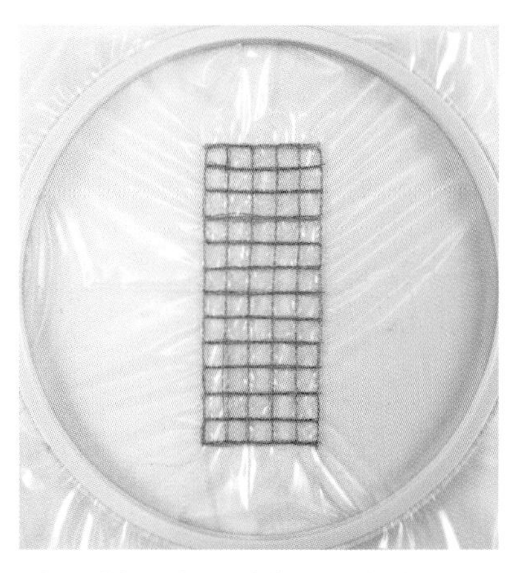

2 Place the stabilizer in the hoop and tighten. Stitch along the drawn pattern lines multiple times. If you want a very delicate lace grid, stitch each line two or three times. For a heavier grid, stitch three to five times along each line. Finish with a zigzag stitch if desired.

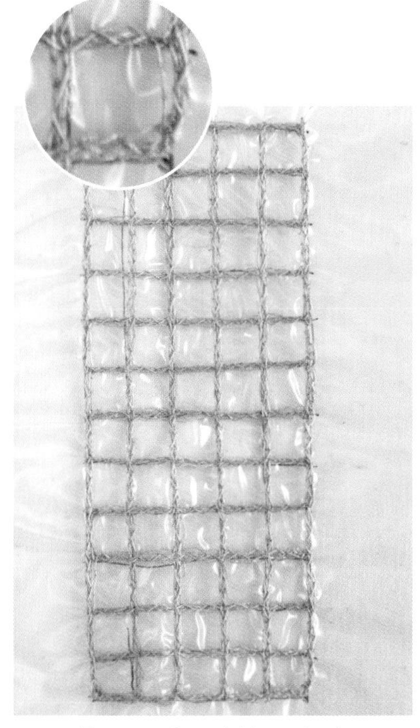

3 Remove the grid from the hoop and trim the stabilizer fairly close to the stitching. To dissolve the stabilizer, either immerse the grid in a bowl of warm water changing the water as necessary or rinse under running water. Block if needed.

tips

- Water soluble stabilizer is available in different weights. Refer to stabilizers on page 20. All of the stabilizers can be layered and are interchangeable. When using light weight stabilizers it is best to begin with two layers.
- You can control the amount of stabilizer rinsed out of your grid. Depending on the final use of the grid you may want it to be soft or firm. Rinsing very briefly will retain much of the stabilizer in the threads and the piece will be firm. Soaking and rinsing will release most of the stabilizer and result in a very soft piece.
- Depending on the thread chosen, lace embroidery will sometimes shrink when the stabilizer is dissolved. If retention of a specific shape is desired, block it by pinning the grid to a padded surface or Styrofoam block as it dries.
- If a thicker, more solid line of thread is desired you can finish the stitched lines with a zigzag stitch to bundle the loose threads together. After you have stitched each line the desired number of times, change your machine stitch to a zigzag. Choose a narrow width. Stitch over each line of the design again.
- Don't worry if you do not stitch directly on the drawn pattern or stitch outside of the desired area. Stray threads may be trimmed after the stabilizer is removed.

thread lace grid

Fibers such as thread, Angelina®, and small bits of fabric or paper can be encased inside a FILLED THREAD GRID.

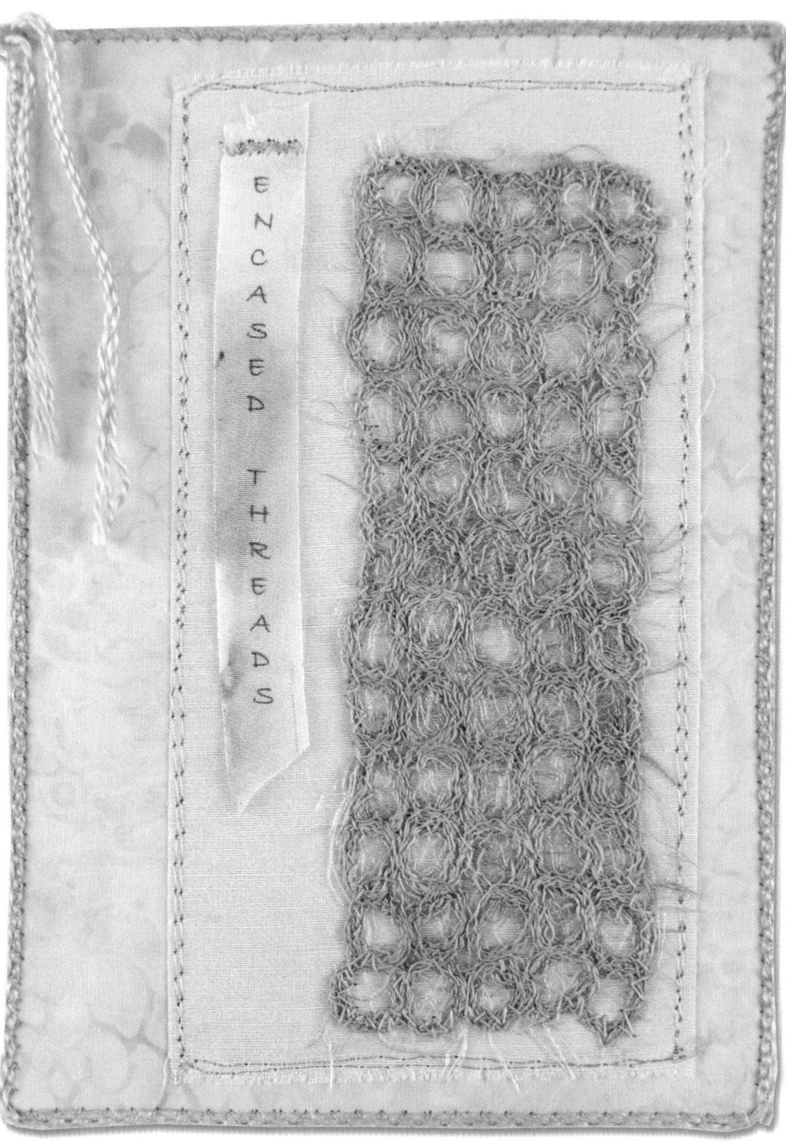

Edge finish: silk cord couched with a zigzag stitch

- In some applications, both sides of your thread lace grid will be visible. Choose the bobbin thread with this in mind. Try a contrasting color thread to add a secondary color or a similar color to add color depth.

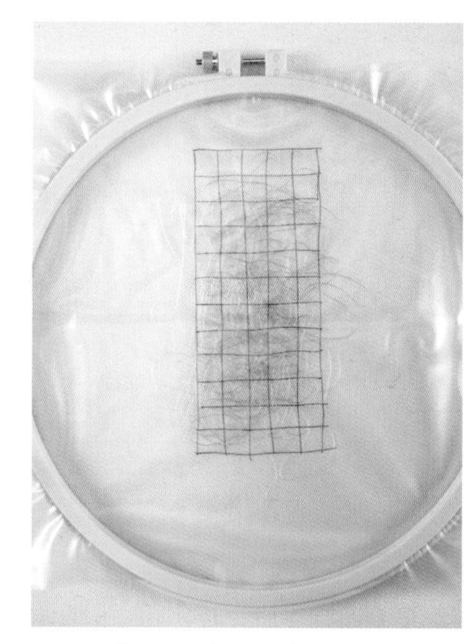

1 Cut two pieces of water-soluble stabilizer slightly larger than your embroidery hoop. If desired, draw a pattern or grid on one layer of the stabilizer with a pen or marker. Lay fibers to be encased on the blank piece of stabilizer. Place the second piece of stabilizer with the drawn design on top of the fibers and place in the hoop and tighten.

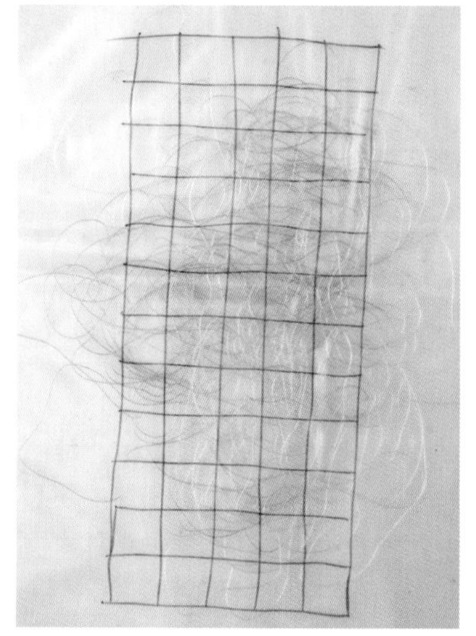

2 Follow steps 2-3 in Basic Thread Lace Grid on page 111 to stitch. In our example we used the grid structure with garnet stitching rather than a straight line.

thread lace grid

THREAD GRIDS WITH PROGRAMMED DECORATIVE STITCHES result in a unique organic lace structure. In this ethereal lace technique the bobbin thread will show so choose a matching, coordinating, or contrasting color rather than a light weight bobbin thread.

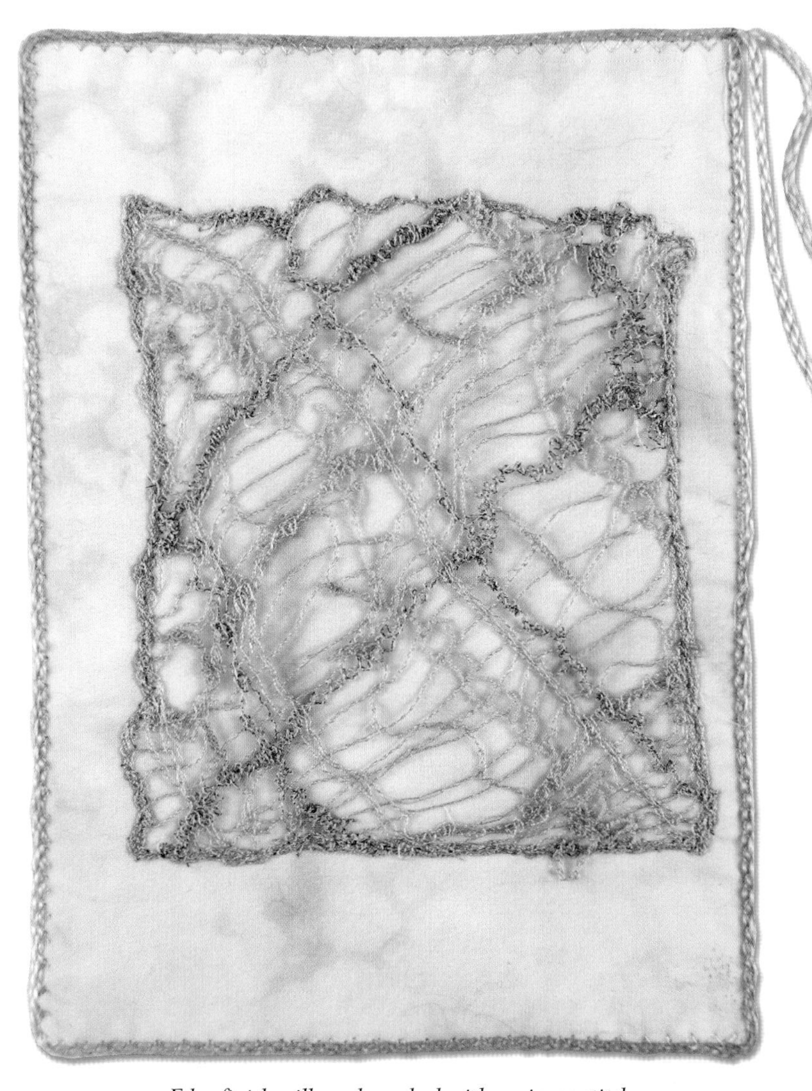

Edge finish: silk cord couched with a zigzag stitch

1 Draw a square or rectangle on heavy duty water soluble stabilizer and secure in an embroidery hoop. Stitch from edge to edge with a free motion straight stitch. Stitch fairly densely, between 1/8" to 1/4" apart. Our sample was stitched on the diagonal for interest.

2 Choose different programmed decorative stitches and stitch in the opposite direction of the straight stitch lines. Stitch a few stitches in the same direction as the straight stitches. Change threads as desired. In our sample we used four different colors of coordinating thread. When you are happy with the stitching trim the stabilizer close to the stitching and rinse out. Block if desired.

Try this

- Change the length or width of the stitches while sewing.
- Use the reverse button to increase the stitch density.
- Stitch a programmed stitch with the machine set up for free motion.
- Try different stitches.

thread lace grid

Use a thread grid to CREATE A THREAD WINDOW. We made a diamond paned window, but you can use this same technique for any manner of cut out shapes in your art.

1 Draw a diamond grid on heavy weight water-soluble stabilizer. Place in an embroidery hoop and set your machine for free motion stitching. Use the same thread in the top and bobbin. Stitch each line three times and finish each line with a zigzag stitch. Trim stabilizer close to the stitching and rinse under running water. Do not rinse all the stabilizer out. For this technique you want the thread lace to be stiff.

Edge finish: double row of crochet weight yarn couched on with a zigzag stitch

2 Trace a window shape on a piece of freezer paper. Cut out and iron onto the fabric where desired. Trace around the window. Remove the freezer paper. Layer the fabric with batting and stitch on the traced line.

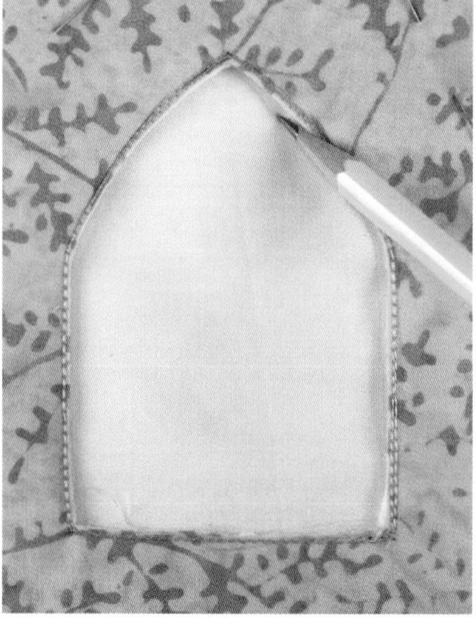

3 Cut out the interior window shape. Lay the fabric with the window cut out on the backing fabric. Trace the window on the backing fabric and cut it out. Sandwich the lace grid between the front and backing fabrics. Use pins to secure. Stitch around the edge of the opening and quilt as desired.

Frog Princess

The frog princess rules her pond in velvet and a free form thread lace skirt. Flowers gathered from the shore decorate her skirt and bodice.

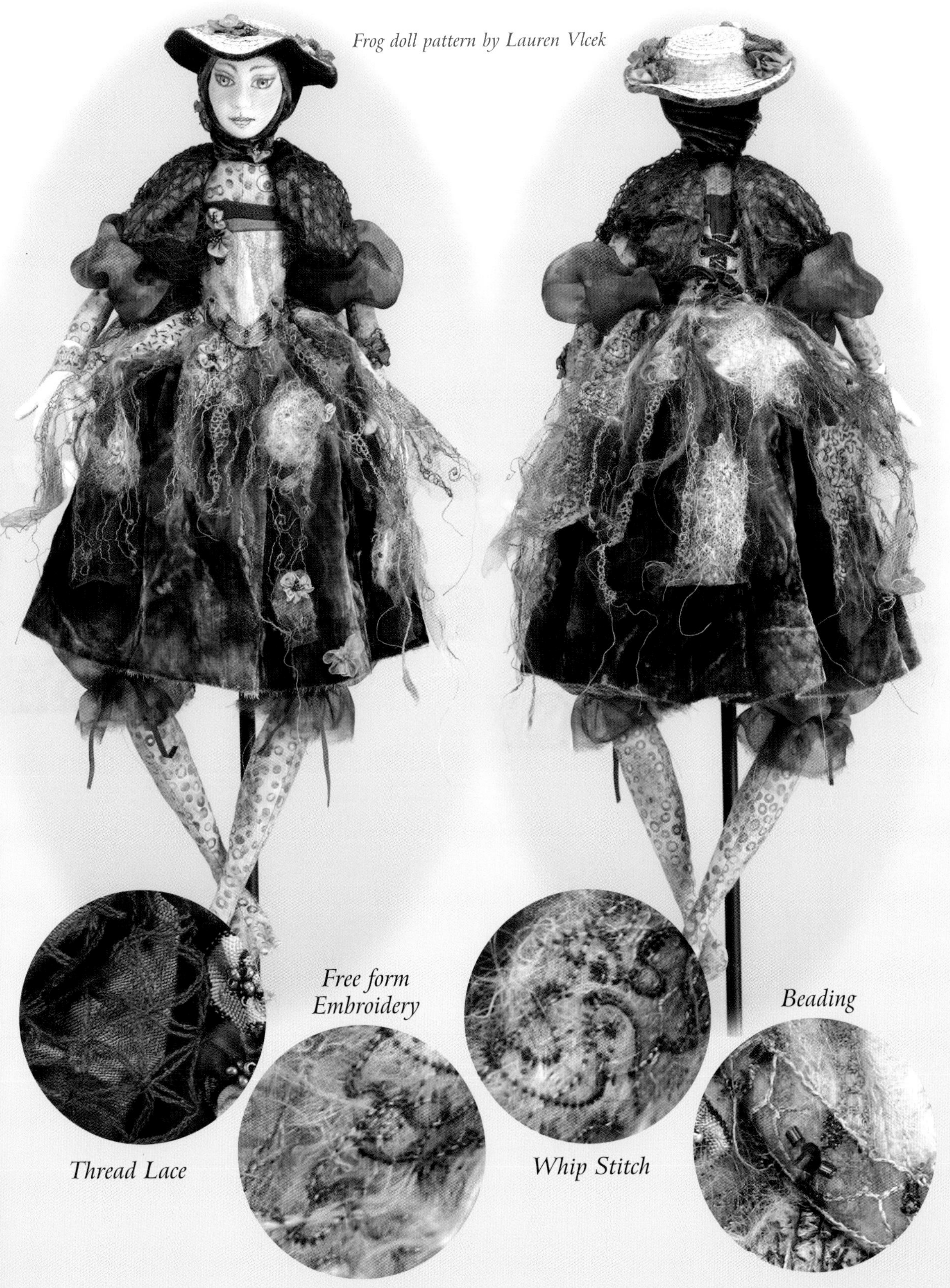

Frog doll pattern by Lauren Vlcek

Thread Lace

Free form Embroidery

Whip Stitch

Beading

thread lace elements

Thread lace motifs are thread painted embroideries created completely out of thread and sometimes tulle or netting. Embroidery motifs may be appliquéd to a surface fabric or used as free standing decorative elements in jewelry and embellishment for clothing, home décor, or quilts. They are created in a similar manner to thread lace grids and use thread painting techniques. Adding a layer of tulle or netting to a layer of water-soluble stabilizer eliminates the need for a securing grid and allows the stitch lines to follow the shape of the piece.

Edge finish: soft yarn couched around edge with zigzag stitch

getting started

Three examples of THREAD LACE LEAVES decorate our Passport page. We appliquéd them to the Passport page by stitching only on the center stem to allow for a dimensional look, but they could be appliquéd on all sides for a more traditional look. The leaves can also be used as a separate dangle and stitched only at the stem.

Machine set-up:
Free motion
drop feed dogs

Tension:
Balanced or slightly loosen upper

Needle:
Embroidery 90/14

Top thread:
Any decorative thread

Bobbin:
Ultra fine or fine polyester

1 Cut a piece of heavy weight water soluble stabilizer and a piece of tulle or fine netting slightly larger than your embroidery hoop. Trace the leaf design below three times on the water-soluble stabilizer or use your own design. Lay the water-soluble stabilizer over the netting and insert in an embroidery hoop.

2 The outline leaf is stitched with a multi-color thread around the outside of the leaf and the veins. Stitch over each line three or four times. Refer to Thread Painting tips on page 69.

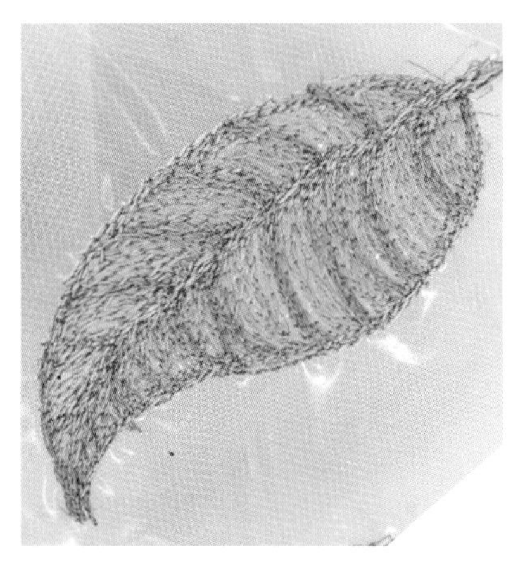

3 The lacey leaf is stitched with a multi-color thread around the outside of the leaf and the veins. Change to a solid green thread and lightly stitch the open areas of the leaf. Follow the curve lines of the veins.

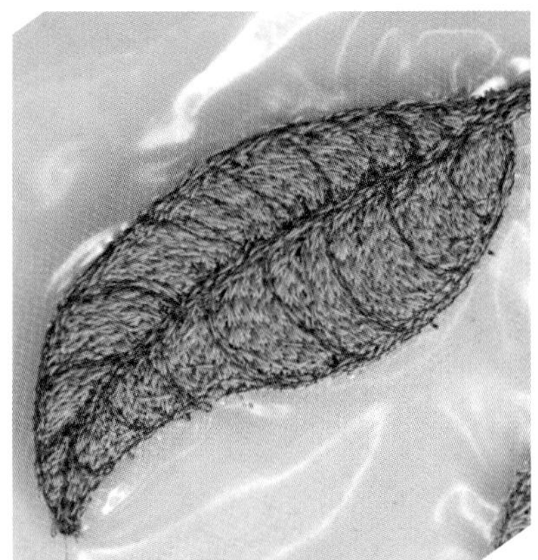

4 The solid leaf is heavily thread painted. The veins and stem are first stitched with a brown thread to provide an outline to follow as well as provide a deeper base color for the veins. Using a solid green thread, stitch the spaces between the veins until they are filled in fairly well. Change to a multi-color green or different shade of green and stitch lightly on top of the solid green. Most of the brown will be covered by this point. Finally, use a dark green thread to highlight the leaf veins and outline. Trim each leaf close to the stitching and dissolve the stabilizer in water. Trim any remaining netting or stray threads.

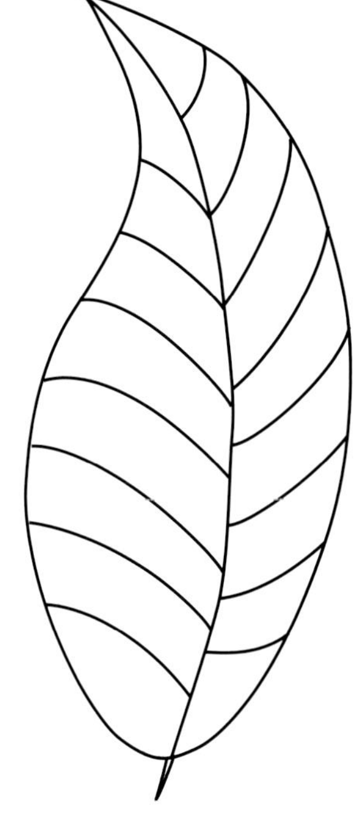

free form embroidery

Specialty water-soluble stabilizers with an adhesive backing are used to create free form embroideries. The stabilizers are used to hold wispy fibers, yarns, delicate fabrics, and ribbons securely so they can be stitched by both machine and hand.

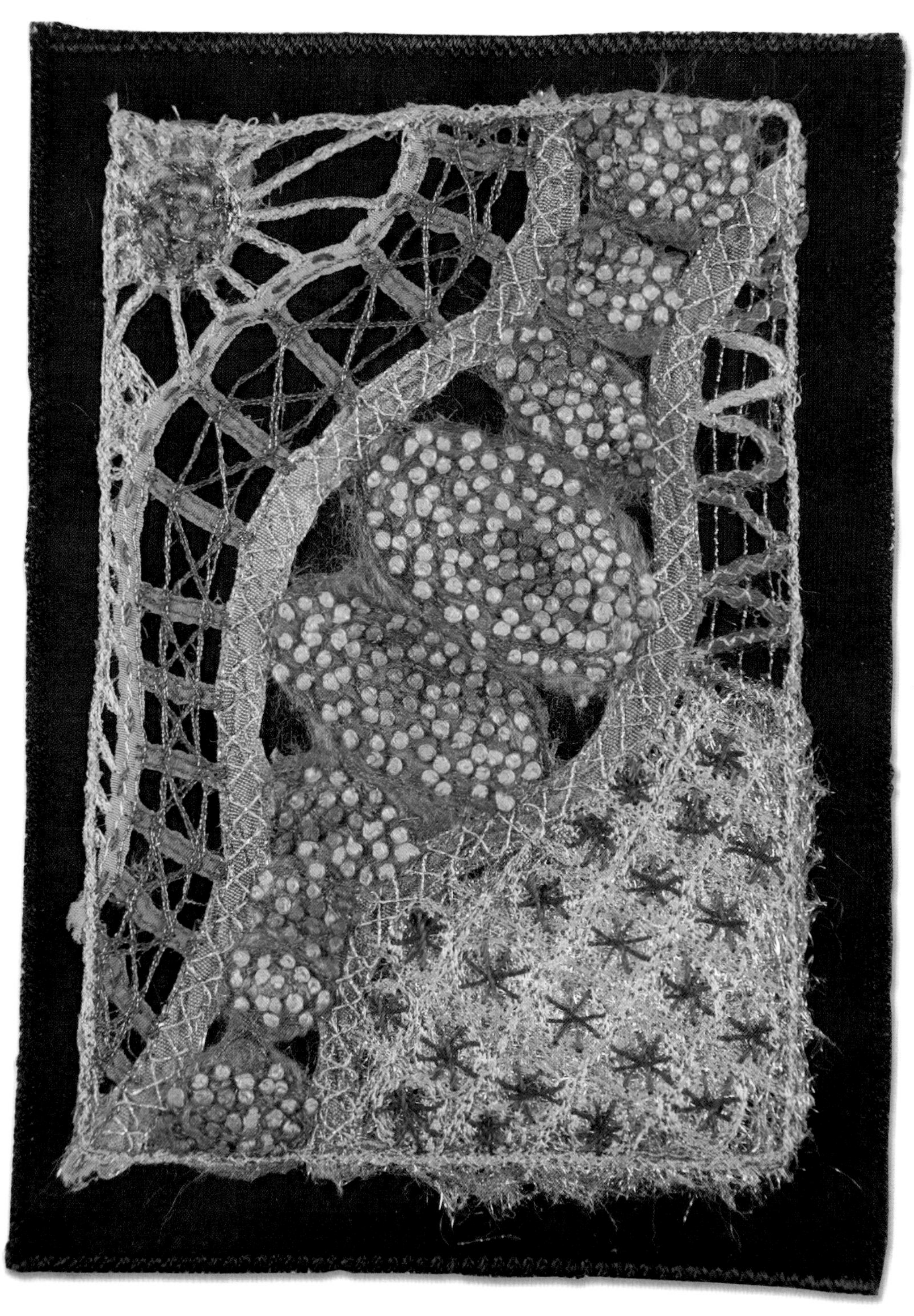

Edge finish: simple zigzag with heavy rayon thread

getting started

This is a unique process where one beginning design can be used to create many variations with simple changes in hand stitching and yarn and fiber textures.

Machine set up:
Free motion
drop feed dogs

Tension:
Balanced or slightly loosen upper

Needle:
Embroidery 90/14

Top thread:
Any decorative rayon, cotton, silk or polyester

Bobbin:
Ultra fine or fine polyester

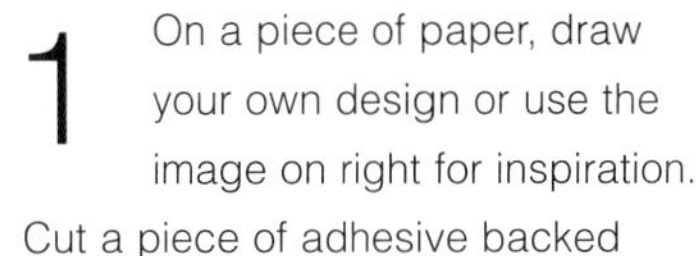

1 On a piece of paper, draw your own design or use the image on right for inspiration. Cut a piece of adhesive backed water-soluble stabilizer larger than your drawn design. Refer to Stabilizers on page 19. Allow 1-1/2" to 2" of stabilizer on all sides of the design. Lay the stabilizer on a firm surface with the paper backed adhesive side up. Begin in one corner and carefully pull off the back to expose the sticky surface. As you expose a corner or side use masking or painter's tape to secure the stabilizer to your work surface. Once you have exposed the entire sheet, slide your drawn design underneath the stabilizer.

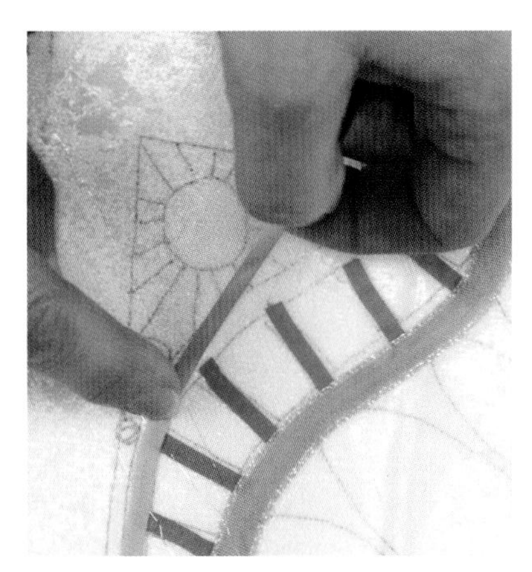

2 Place fiber, yarns, and threads over parts of the drawn design. It is easiest to place the non-fuzzy fibers first. We used rayon and silk ribbons over curved and straight lines, and silk cord over zigzag lines.

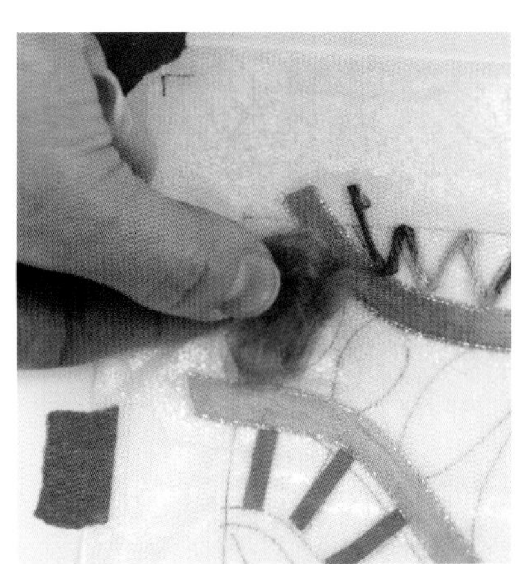

3 Small bunches of silk roving are placed over the large oval shapes and one circular area.

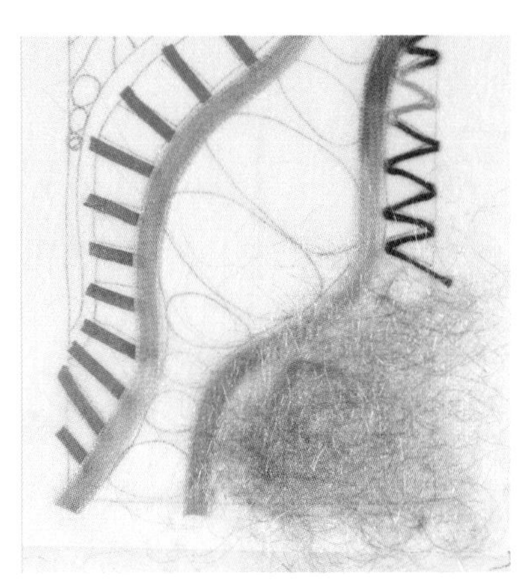

4 Angelina® fibers are laid over the corner circles and grid lines. Additional Angelina® fibers are laid on top of the silk roving.

5 Carefully place a piece of light weight water-soluble film, often referred to as topping or backing, on top of the adhesive water-soluble film and fibers. Press the top layer down firmly to meet the remaining exposed sticky film to secure and create a fiber sandwich. Use a Sharpie® or permanent marker to trace the pattern on the top layer of water soluble stabilizer. Don't worry about tracing the exact lines of the original pattern. One of the fun aspects of this technique is being able to change or adapt the pattern on a whim.

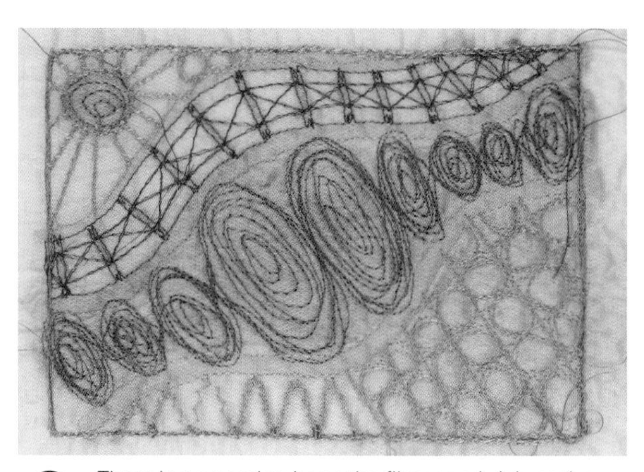

6 There is no need to hoop the fiber sandwich as the two layers of water-soluble stabilizer create a very sturdy stitching surface. Begin by free motion stitching all the areas with loose fibers, yarns, and ribbons. Next, stitch any remaining drawn lines that don't contain yarn or fibers. The same rules and tips for making machine thread lace apply. Refer to Thread Lace Grid tips on page 111. Make sure all the stitching connects on multiple points. The stitching on the loose fibers needs to connect to the edges of the embroidery and ribbons that run along both sides. All elements must connect to all other elements. Holding the embroidery up to the light can help identify areas needing additional stitching. It can look very messy at this point. This photo shows the reverse side of our sample.

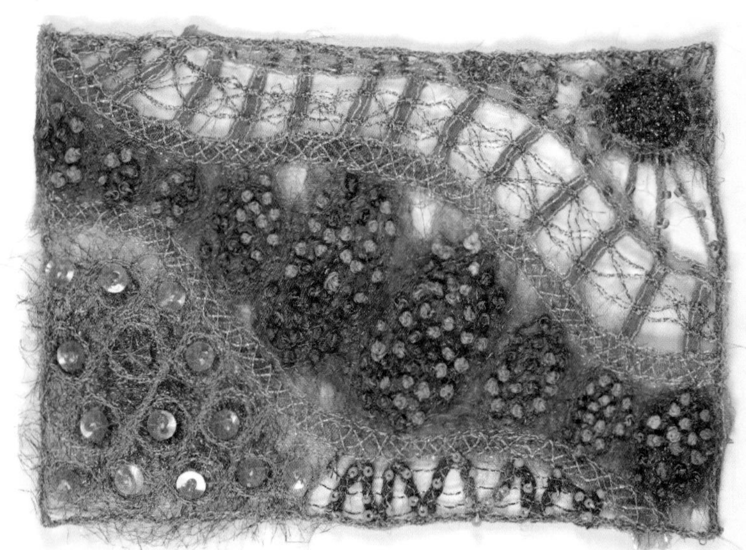

7 After ensuring all areas are linked with machine stitching, hand stitching and beading can be added. It may seem difficult to decide where to start. Simply choose your favorite hand stitch technique and try it out. Our piece began with French knots in the middle of the silk roving ovals. Very simple stitches can give wonderful results. Try adding fibers with couching. Use cross stitch, running stitch, blanket stitch, and other simple embroidery stitches in a variety of thread fibers and weights. Refer to Hand Stitching on page 94 for stitch samples. Chenille needles work well for this technique.

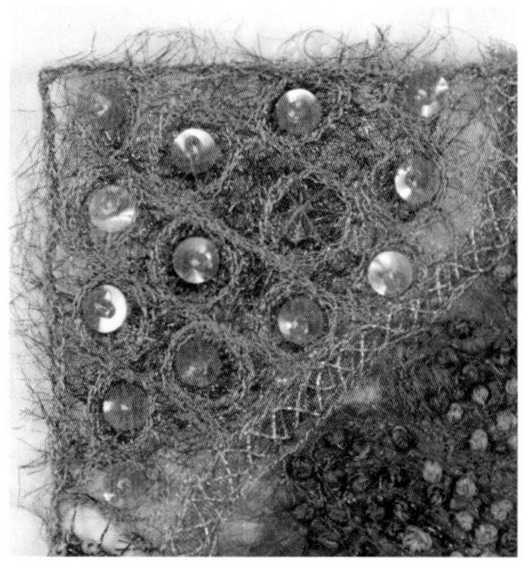

8 When you are satisfied with the amount of hand stitching, trim the water soluble stabilizer and rinse the embroidery in water to dissolve the stabilizer. Block if desired and let dry. Apply the embroidery to your Passport page with a zigzag or hand couching stitch.

free form embroidery heart

The same techniques for free form embroidery can be used to create solid shapes using ethereal silk fibers, paper, and assorted fabrics.

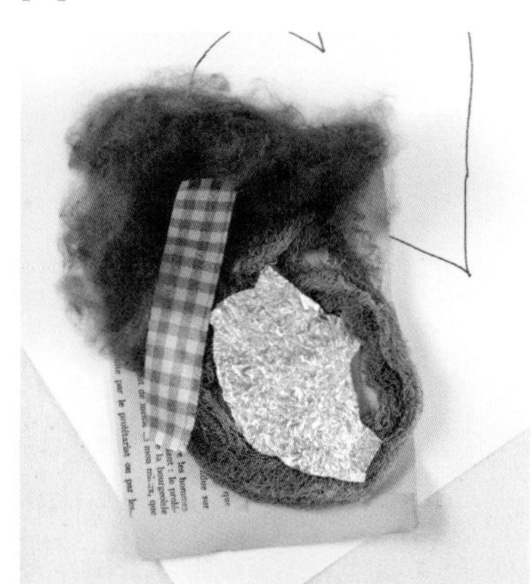

1 Draw a heart shape on a piece of paper. Following step 1 in Free Form Embroidery on page 119, expose the adhesive side of the water-soluble stabilizer and slide the heart pattern under it. Place fibers on the heart shape. We used one layer of silk hankie, bits of vintage paper, chocolate foil bits, dyed cheesecloth, and small squares of silk fabric. The silk fibers are wispy and won't all adhere completely to the water-soluble stabilizer. You will have a somewhat fluffy pile of fibers and fabrics.

Edge finish: silk ribbon couched with a zigzag stitch.

2 Place a piece of light weight water-soluble stabilizer on top of the adhesive stabilizer and heart shaped fibers. Press with your hands to secure the layers. Free motion stitch the surface of the heart. We used a loose grid stitched in a wavy line. Try some of your programmed stitches for fun and added texture. Because the base of the heart has fabrics in addition to the web structure of the silk hankie, the machine stitching need not be as dense or connected as in the free form embroidery.

3 After you have a base of machine stitching, add hand stitching as desired. Try a variety of stitches, threads, and yarns. Ribbon tails were left dangling. Rinse the embroidery in water to dissolve the stabilizer, let dry and apply to a Passport page base with additional stitching.

small thread structure

Fitting a three-dimensional thread structure between two pages can be a little awkward, but it's a good warm-up for creating larger vessels, vases, and bowls.

Edge finish: light zigzag stitch and a couched fiber

getting started

Stitching out a thread grid, considering the inclusions it can contain, and working to mold a shape are all part of the skills you need to go 3-D in a bigger way.

Machine set up:
Free motion
drop feed dogs

Tension:
Balanced or slightly loosen upper

Needle:
Embroidery 90/14

Top thread:
Any decorative thread

Bobbin:
Medium weight polyester or matching

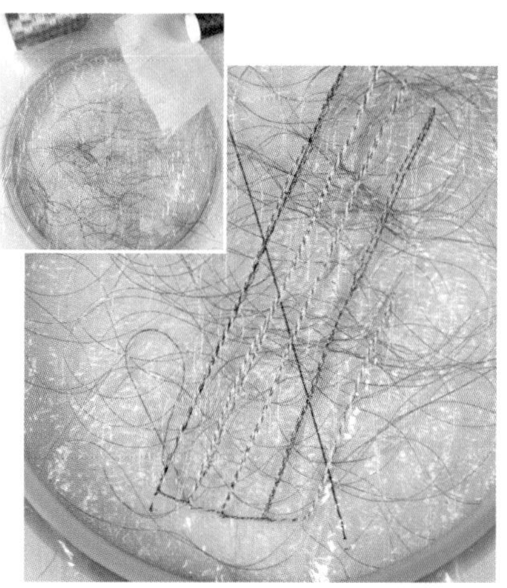

1 Pre-wrap the molding base with plastic wrap. We used a glass bud vase as our base. This wrapping keeps the thread structure from bonding to the vase as it dries. The base should be no bigger than 2" to 3" across.

2 If you want to add bits of fabric or fibers to your structure, capture them between two layers of water soluble stabilizer. Or, follow the instructions for Filled Thread Grids on page 112, but use the entire inside dimension of a 6" embroidery hoop. We added threads and small bits of fabric as our inclusions or filling. The grid above was stitched in a diamond pattern.

3 When the grid is complete, remove it from the hoop and trim away the excess stabilizer. Rinse the grid piece under lukewarm water. Don't completely rinse out the stabilizer. The remaining stabilizer will dry and firm up the thread structure. Squeeze the excess water out of the thread grid. Drape the grid over the pre-wrapped base shape. Make sure the edges are fairly equal in length. Secure in place with a rubber band. The grid will shrink slightly as it dries. Normally, a thread vessel is allowed to dry on its own to avoid distortion. We wanted to flare the edges so we helped the drying process with a heat gun. Lightly dry the edges all around so that they stand away from the mold. Once the edges are dry, move the heat gun around the remainder of the grid. Remove the rubber band and dry this area as well. For added strength, you may wish to add a coat or two of matte spray sealant when the structure is dry. NOTE: Certain threads or inclusions like Angelina® may not respond well to the heat drying. These may need to air-dry with the edges propped in place.

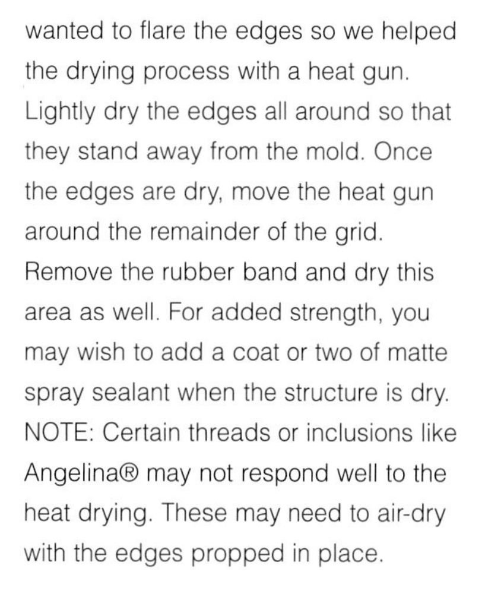

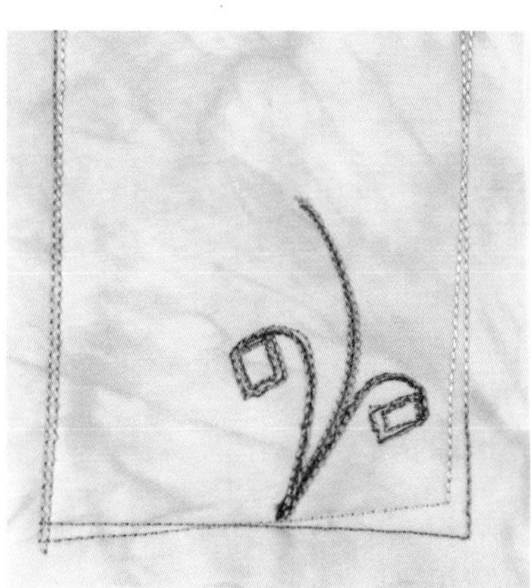

4 Prepare the Passport page by stitching around the perimeter with a few "wonky" rectangles in a heavy weight thread. Estimate the placement of the stem by holding the structure in place over the base. Mark the stem line if desired. We free motion stitched the stem and square leaves in blue and then in green.

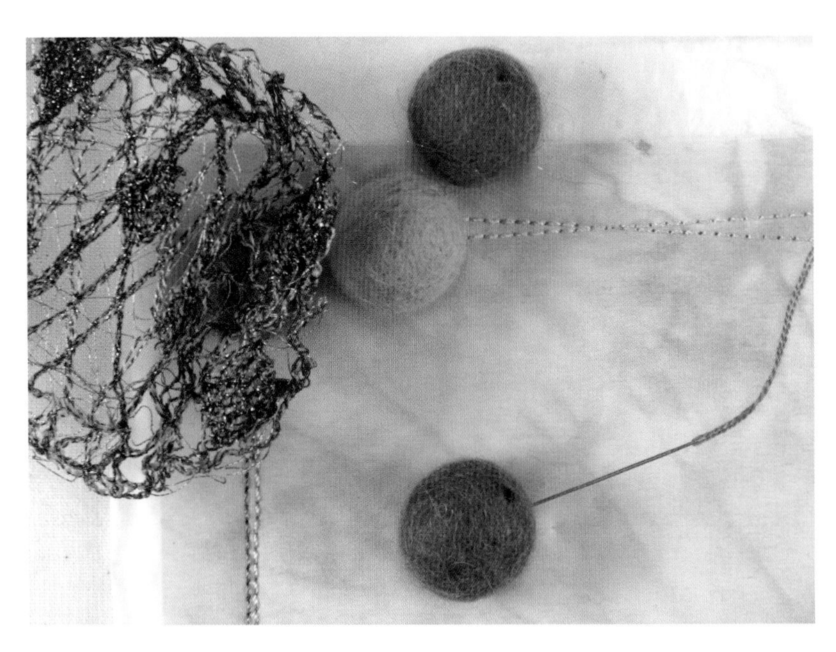

5 To keep the structure from being flattened within the Passport, a felt bead was added as the flower center for support. The bead was hand tacked in place with a few stitches through the center. Do not snip the thread. Lay the thread structure over the bead after tacking it down and continue stitching to tack the structure in place. A few stitches in the center will secure the grid to the bead, while a few stitches around the flared perimeter will secure it to the base fabric.

Try this

- Add silk bits, torn pieces of Angelina® or a greater amount or variety of thread bits to add interest to the composition. It will also allow more space between the grid lines.
- A zigzag stitch can be used, but it must be stitched over a straight stitch so it does not open up and fall apart when the stabilizer is removed.
- After the grid is completed, you can add a few intermittent fills if you wish. Stitch around the grid lines, stopping at random grid spaces to fill with stitching. Stitch horizontally and then vertically between the outline stitches. Be sure to cross over the outline stitches to secure the fill within the grid.
- Try a variety of thread structure shapes.

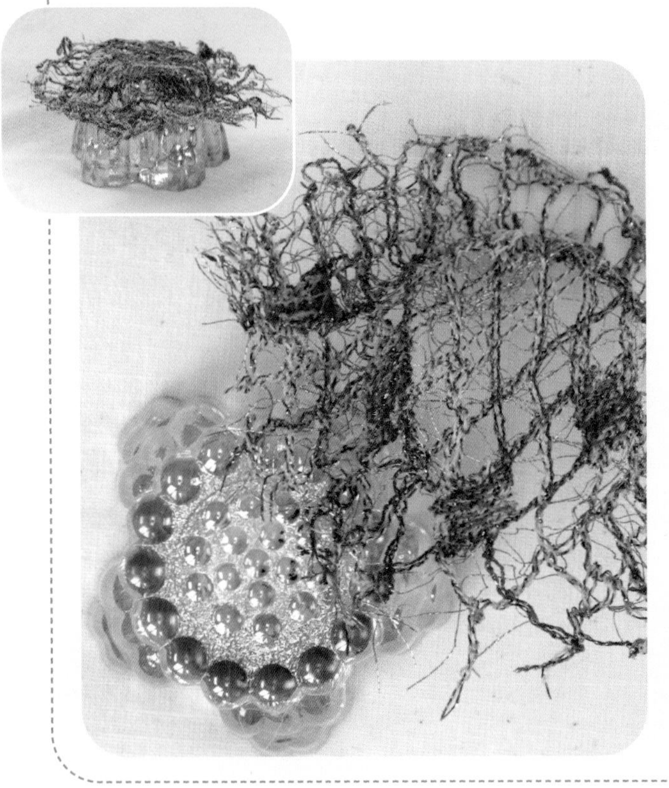

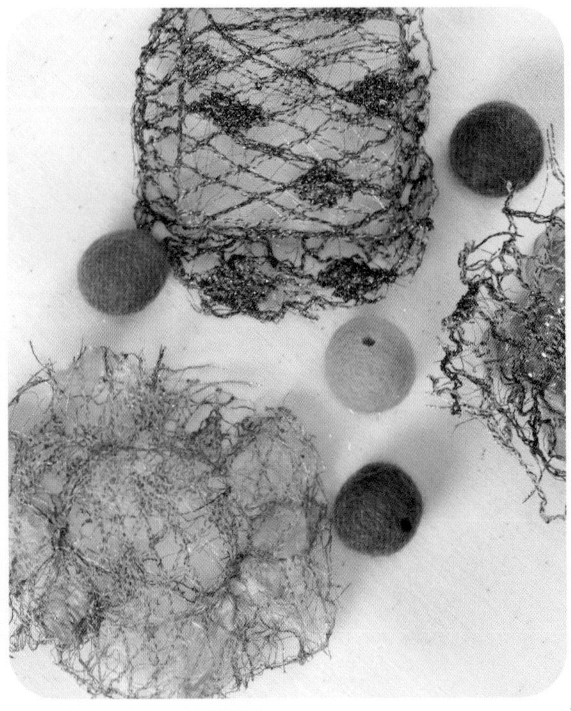

Cross My Heart

This thread structure was formed over a heart-shaped glass dish and then randomly stitched with wavy lines of metallic thread before being placed in a frame. A ribbon is woven through the structure and tied in a bow with the ends left loose to float from the frame.

Thread Structure

Thread Painted Appliqué

Cross Stitch

Ribbon

the techniques

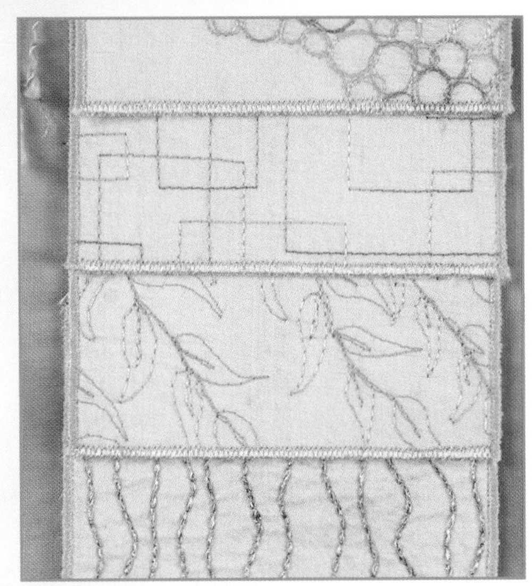

make your own fabric

make your own lace

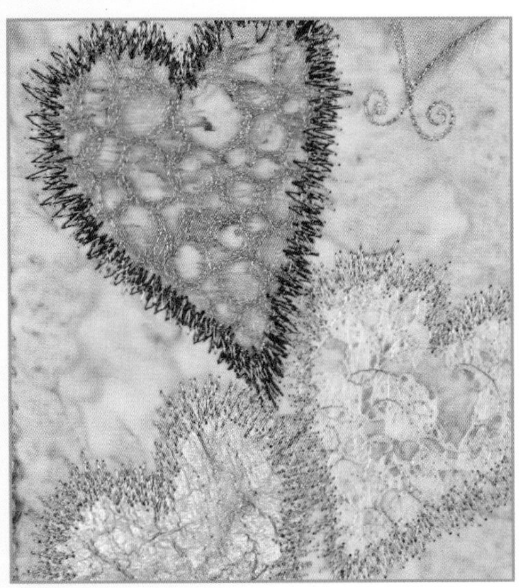

thread as a resist

machine wrapped cords & sticks

sashiko inspired stitching

tassels & fringe

Thread as Ornament

Stitch has the ability to transform a plain fabric into something spectacular when you make your own fabric, add color, pattern, texture, and style. 'Make your own fabric' creations can be used in quilts, home décor, and wonderful accents for garments.
Create fascinating embellishments using thread as a resist with paper, Lutradur®, Tyvek®, Textiva®, and chiffon. Play with ribbons, explore your programmed stitches and stitch on bamboo skewers and cords to make custom dimensional ornaments. This section provides a treasure trove of ideas to get your creative juices flowing.

make your own fabric

Most of our samples are created using programmed stitches with the feed dogs up. Some are made with free motion stitch patterns. Try combining free motion with programmed stitches.

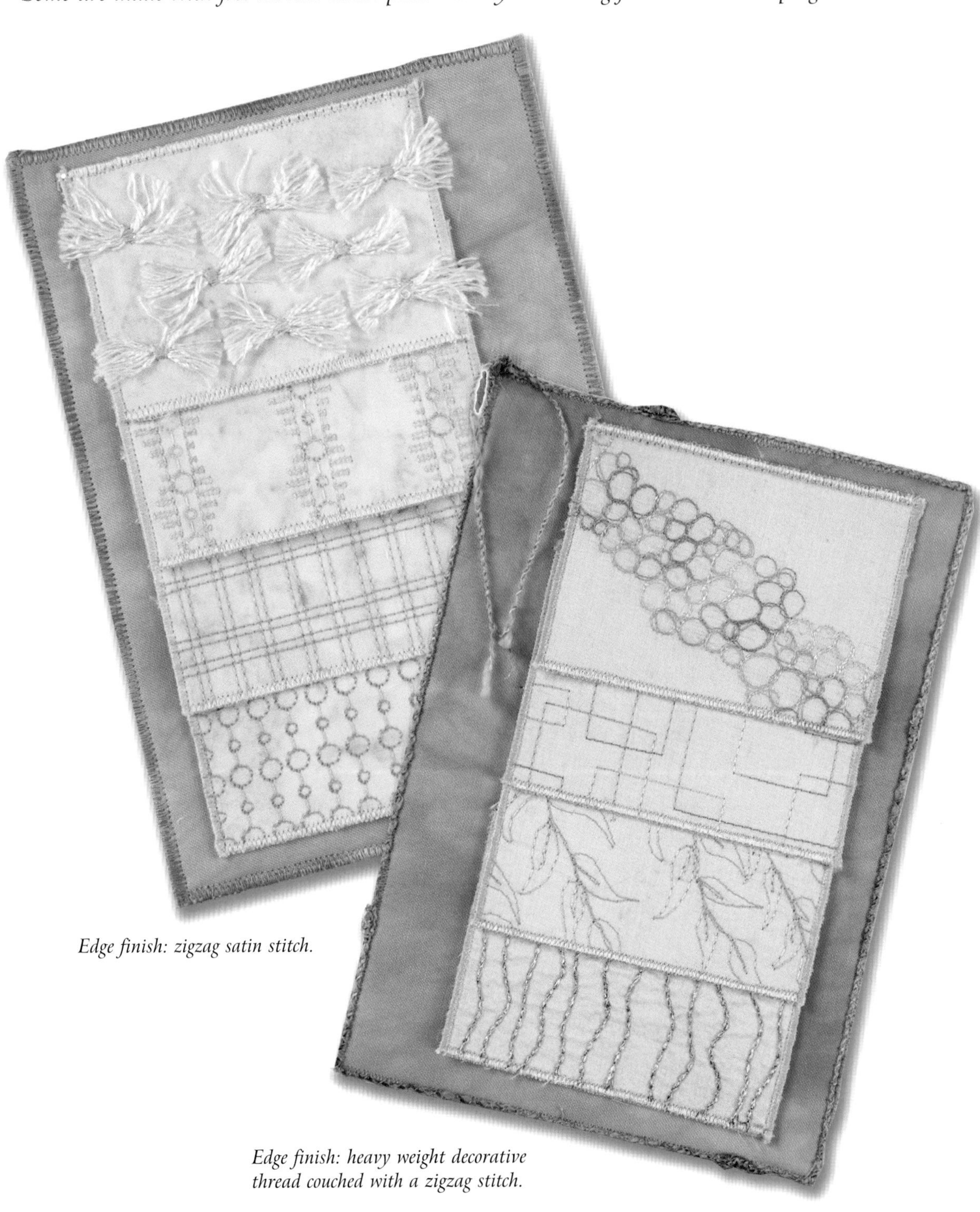

Edge finish: zigzag satin stitch.

Edge finish: heavy weight decorative thread couched with a zigzag stitch.

getting started

Cut your fabric slightly larger than your desired finished size to allow for shrinkage. Our samples were cut 3-1/2" x 4-1/2" and trimmed to 2-1/2" x 3-1/2" after stitching was done. Back each piece with a light weight stabilizer. Choose stitch ideas from our samples and steps or create your own.

Machine set up:
Normal

Needle:
Embroidery 90/14

Tension:
Balanced

Top thread:
Any cotton, polyester, silk, or rayon

Bobbin:
Light weight or matching

1 TUFTS: Use a pencil or disappearing marker to mark straight guide lines. Lay multiple strands of yarn on your fabric. Choose a zigzag stitch (satin stitch length) and tack the strands at regular intervals.

2 Turn the fabric 90-degrees. Using a zigzag stitch tack the strands in this direction.

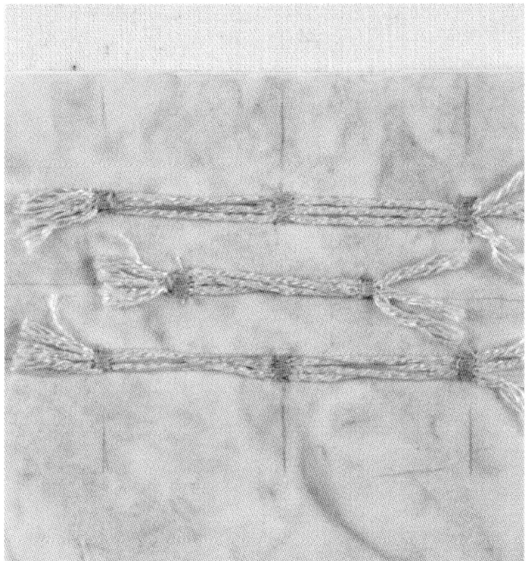

3 After all the strands are tacked down, snip the yarns between the tack stitches and fluff the yarns.

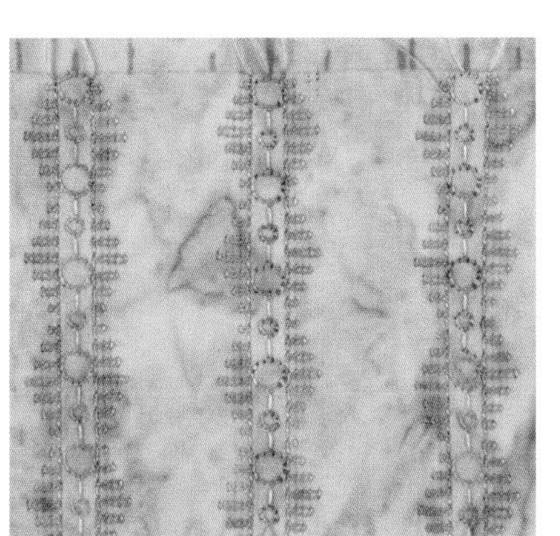

PATTERNED ROWS:
Choose stitches from your Programmed Stitch Guide that will combine well to create a new design. Refer to Programmed Stitch Guide on page 23. Stitch the middle pattern stitch, then stitch the side stitch design. There is some trial and error in this process to discover favorite unique stitch combinations.

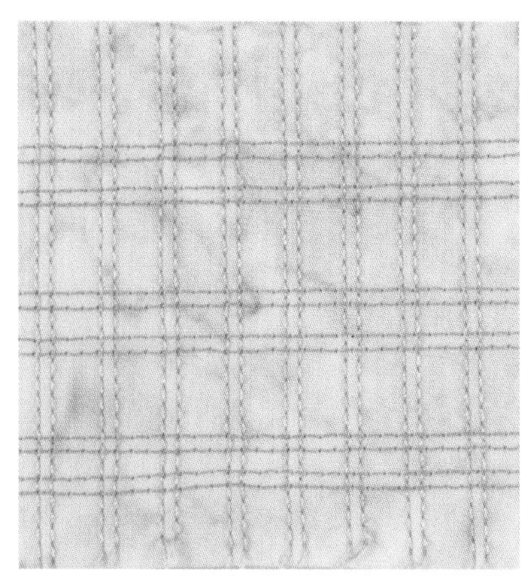

TWIN NEEDLE GRID: A twin needle is used to create an asymmetrical grid. Our sample used only one color of thread. Try using different thread colors or a variegated thread.

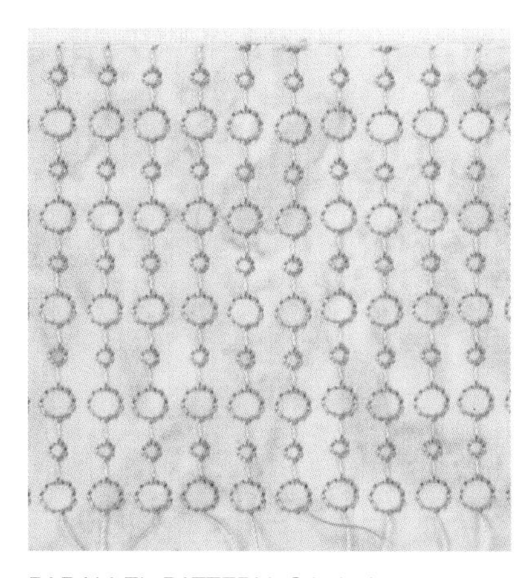

PARALLEL PATTERN: Stitch the same pattern immediately next to the previous one to create fun designs.

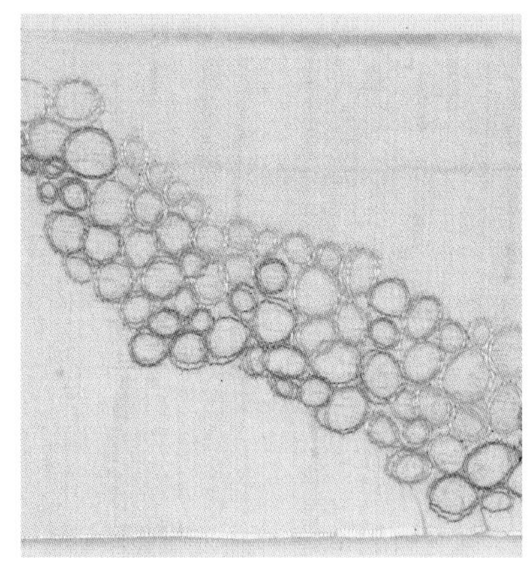

FREE MOTION BUBBLES: Use the garnet stitch and a variegated thread to create a river of bubbles on silk fabric.

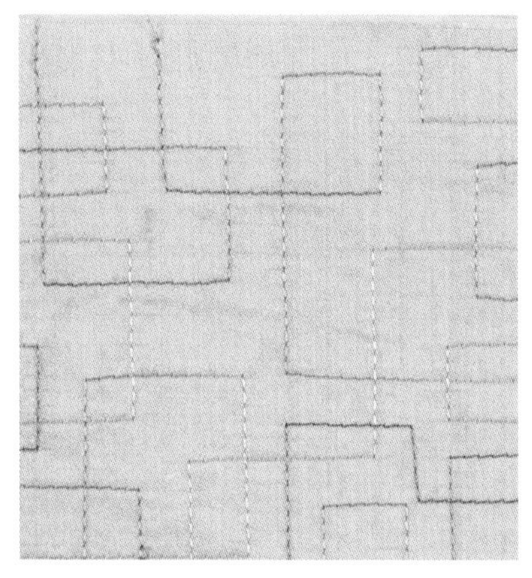

BOX PIVOTS: These may be made with the feed dogs up or down. Pivot to change directions to form boxes. Use your needle down option or put your needle down into the fabric at each pivot point.

FREE MOTION PLAY: Our sample features leaves cascading down silk fabric. Choose your favorite free motion stitch motif to create your own couture fabric.

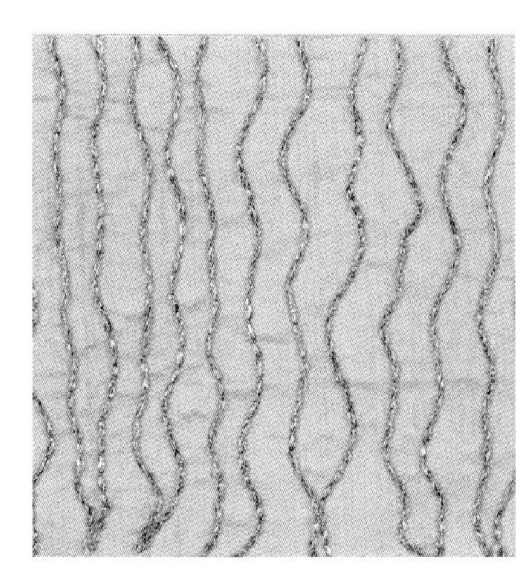

BOBBIN RIBBONS: Metallic yarn used in the bobbin is stitched in a simple undulating line to create a fun playful fabric.

Trim the fabric samples you have just finished to 2-1/2" x 3-1/2". Fuse a plain fabric to the back. Zigzag stitch the edges of your created samples.
Layer a fabric page base with batting and light weight stabilizer. Attach your fabric samples with a straight stitch.

Designer's Workshop

These trading cards show free motion stitching, doodle stitching, and a bead embellishment. Sign your name on the back to personalize a tiny gift of art.

make your own fabric

Stitching on sheer and delicate fabrics is easy when you use heat or water-soluble stabilizer. Play with your favorite free motion patterns to create one of a kind fabric for garments, home décor, art, or quilting.

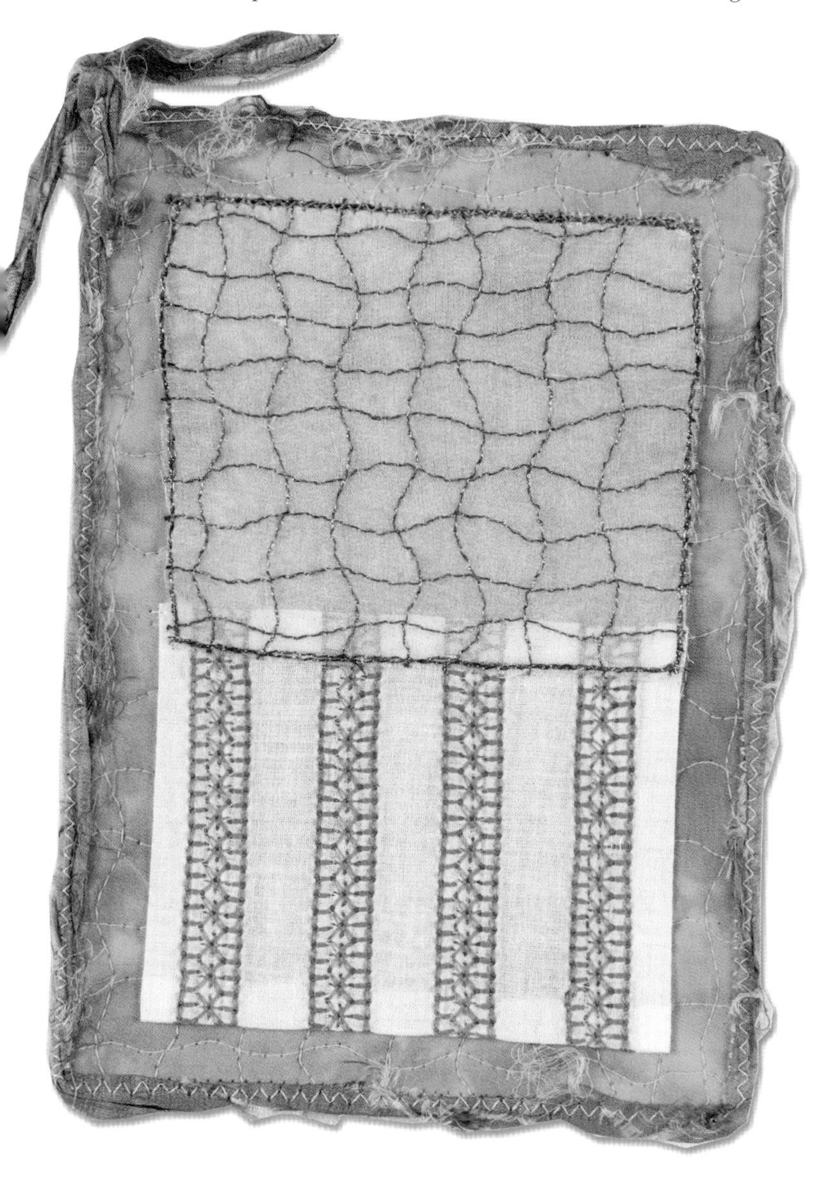

Edge finish: ripped and frayed piece of silk fabric lightly twisted while being couched with a zigzag stitch.

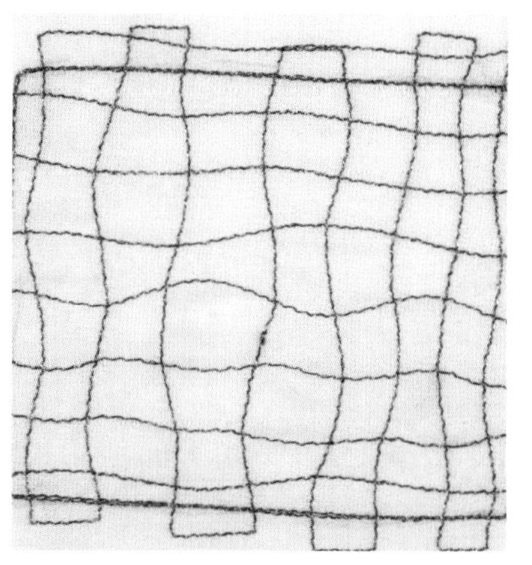

SHEER SILK ORGANZA: When stitching on sheer fabrics use the same thread in the top and bobbin if both sides will be seen in the finished piece. Layer water-soluble or heat away stabilizer under your fabric. Pin in place or use a temporary spray adhesive.
Free motion stitch your desired pattern then remove the stabilizer following manufacturer's instructions.

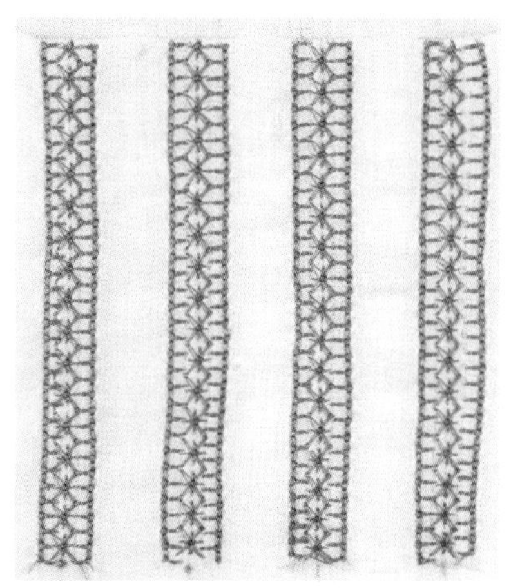

HEIRLOOM STITCHES: Use a fine cotton batiste fabric. Apply spray starch to the fabric to provide additional stabilization. Layer the fabric over a heat away stabilizer. Water-soluble stabilizers can be used, but the final fabric will have a slightly gathered appearance between the rows of stitching. Insert a wing needle in your machine. Use a fine rayon or polyester thread. Thick threads do not work well with this technique. Choose programmed stitches that penetrate the same hole multiple times to create the traditional look of heirloom stitch work. Stitch rows of patterned stitches as desired.

make your own lace

Soft, delicate French lace drapes beautifully and makes an exquisite addition to fashion and home décor. Achieve similar results for a fraction of the cost with free motion stitching over tulle.

Edge finish: zigzag stitch to hold band of tulle between stabilizer and page.

Machine set up:
Free motion
drop feed dogs

Tension:
Balanced

Needle:
Embroidery sized to top thread

Top threads:
Any decorative thread

Bobbin:
Fine

Designer's Workshop

In addition to creating lace, stitched tulle also makes a great overlay. Use a stabilizer that easily washes away to maintain the soft feel of the tulle.

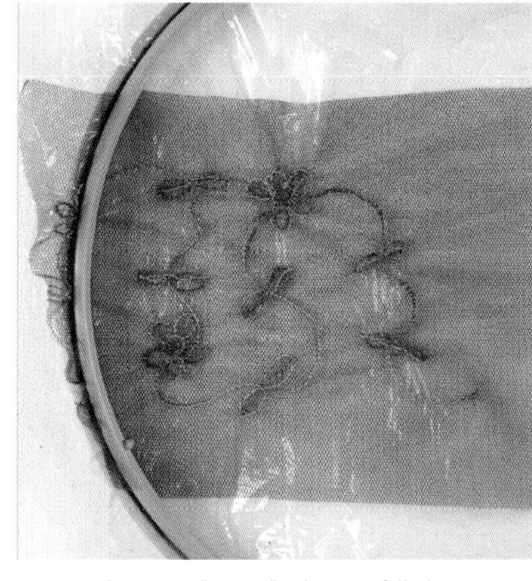

1 Lay a 6" x 15" piece of light weight soft tulle over a light weight water-soluble stabilizer and secure in an embroidery hoop. Free motion stitch a flower design on the lace, in an approximate 4" x 6" band. Leave at least 1-1/4" along one long side unstitched.

2 Trim away excess stabilizer from the long edge so it can be folded and stitched. Fold a 1/2" hem along this edge and press if needed. Be careful not to overheat the tulle. Finish using a fine thread and an invisible hem stitch to create a shell tuck. Refer to the Stitch Dictionary on page 22.

Stitch a straight stitch line left of the shell tuck to secure the raw edge to the back of the fabric.

Trim the two outside band edges at slight angles to form the 'skirt.' Turn these edges under and stitch a 1/4" hem. Leave the bottom edge raw.

Rinse away the stabilizer and let dry. Be sure all residue is removed to maintain a soft drape.

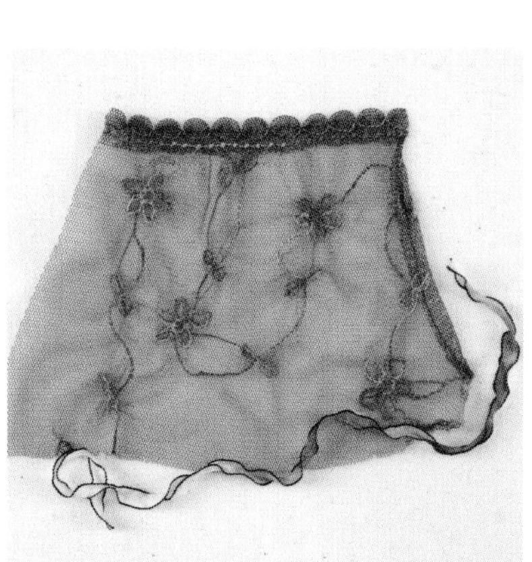

3 Hand tack a bit of stitched ribbon or trim to the bottom edge of the skirt to conceal the raw edge.

4 Stamp an image on the center of the page. Add stabilizer behind the page. Take a few gathering stitches at the top of the skirt and tack it at the mannequin's waistband. Tack in a few more spots to secure the lace to the page.

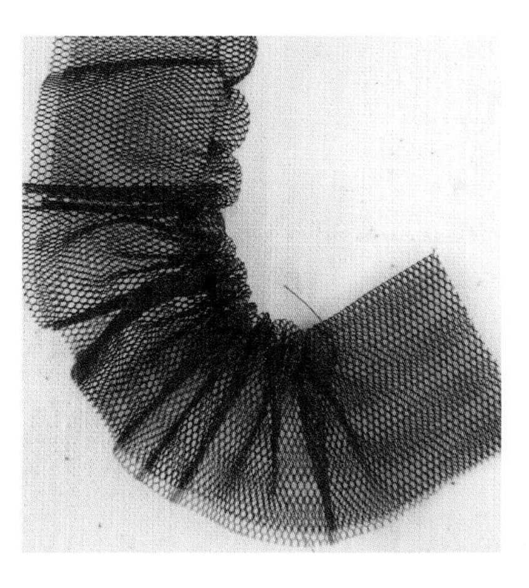

5 Repeat the shell tuck stitch on another band of tulle and tuck it into the outside edge of the Passport page. Add a few rows of metallic thread for embellishment.

thread as a resist on heat distressed fabrics

The heat gun is ideal for distressing, burning or melting away heat-sensitive fabrics such as sheer synthetics, acrylic felt, Tyvek®, Textiva®, or Lutradur®. When these fabrics are stitched and then heated, the fabric breaks down or changes where it is not stitched. The thread acts as a resist to this disintegration of the fibers, leaving behind delicate filaments of texture.

Edge finish: lightly stitched with a zigzag and yarn couched around the edges.

Edge finish: sheer ribbon folded and sewn around the edges with a zigzag stitch

getting started

We worked with an acrylic felt, not a wool or wool blend felt. Before stitching several rows of one pattern, do a trial run with a sampler of different threads and stitches to discover your favorites. There is no one right stitch pattern. Use what you have and what you like.

Machine set-up:
Normal

Tension:
Balanced

Needle:
Universal or Embroidery 90/14

Top thread:
Any decorative or cotton thread

Bobbin:
Medium weight

1 Fill a 6" x 9" acrylic felt piece with rows of decorative stitching. Use the sewing foot as a guide to equally space the rows.

2 Working on the back of a heat resistant surface, use a skewer to move the fabric and hold the felt securely beneath the heat gun. The heat gun should be approximately 3" above the felt and moved in a slow swirling motion to avoid over-heating any area. The felt will distress quickly.

3 When the felt is distressed to your liking, dry brush the highlights. This is the step that makes the ordinary felt appear extraordinary. Load a paintbrush with a small amount of acrylic paint. Brush the paint across the texture of the felt repeatedly. Build-up the color, but do not wet the brush or dilute the paint. If the paint is too wet the felt will soak it up and become saturated. Add more than one paint color if desired.

4 Separate the rows of stitching by gently tearing. If you need to use scissors, take short little snips to leave a rough edge. Apply more paint if desired.

5 Weave the rows of stitching together on top of the background fabric. Stitch in place with a metallic thread. NOTE: The horizontal strips in the photo were flipped over to demonstrate the weaving pattern. Keep all of your pieces facing up.

- Heat distressed fabrics are for decorative use only. They are too fragile to wash.
- Synthetic fibers do not all distress the same. Some will melt, so experimenting is necessary.
- The air from the heat gun will become hot quickly. Pay attention and be prepared to pull the heat away to avoid over-distressing the fabric.
- Work in a well-ventilated area or outdoors.

thread as a resist

Use sheer synthetic fibers, Lutradur® & Tyvek® to create the Passport page on page 134.

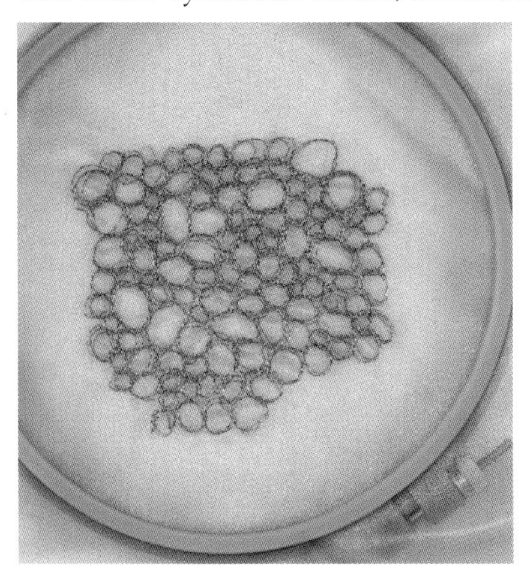

1 Secure a piece of sheer synthetic fabric in an embroidery hoop. The fabric should be taut. The bobbin thread color will be visible through the sheer fabric. Fill the fabric with free motion stitches. We used a garnet stitch in a metallic thread. Set aside.

2 Lightly color a small piece of Lutradur® with spray ink, Glimmer Mist, or a diluted acrylic paint wash. The Lutradur® does not need to be hooped for stitching. It is a fairly firm, non-woven material. We used a free motion looping stitch in a variegated thread.

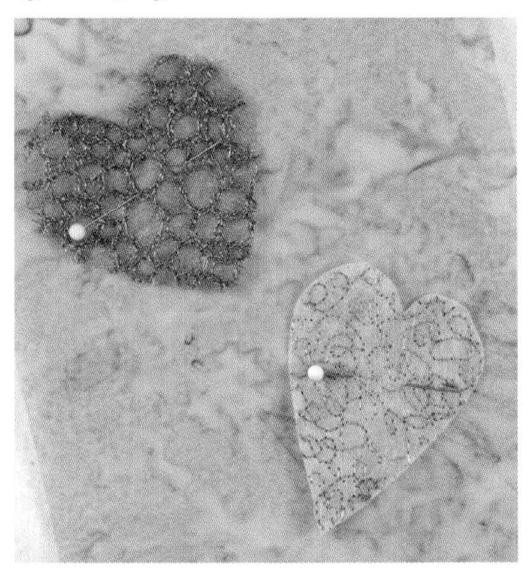

3 Cut a heart shape out of the sheer fabric and the Lutradur®. Lightly appliqué in place on the page. More stitching will be added later.

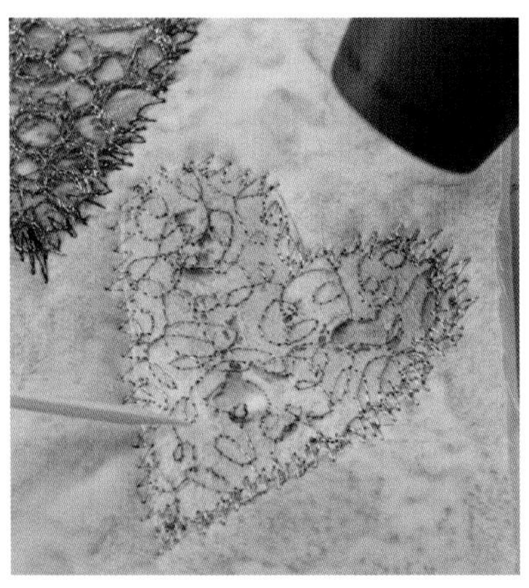

4 Use the back of a cookie sheet and a skewer and lightly distress the two appliqués with the heat gun. Don't overheat.

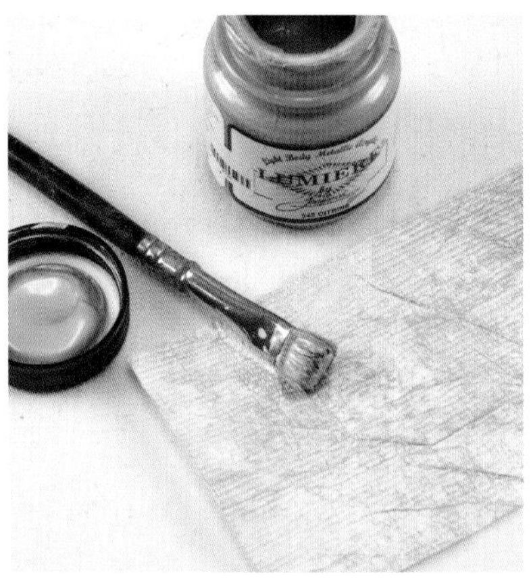

5 Lightly paint one side of Tyvek® with diluted acrylic paint.

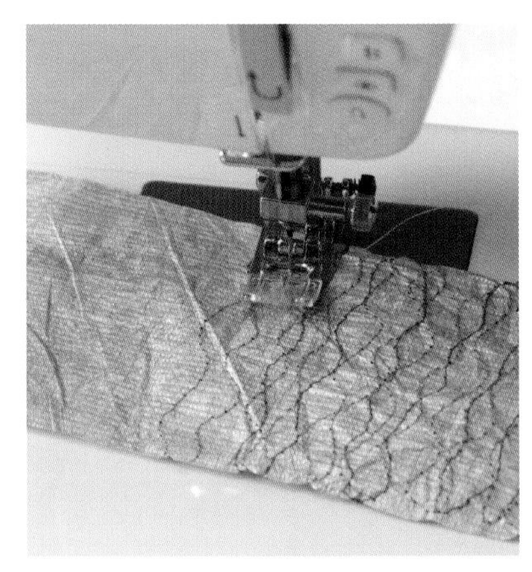

6 Stitch wavy lines on the Tyvek® in one or two threads.

7 Heat the painted side of the Tyvek® until it begins to distort. Turn it over and repeat on the unpainted side. Notice the different ways the fabric reacts. When you are satisfied with the degree of distortion, cut out a heart shape. Appliqué the Tyvek® heart to the page. Stitch around each of the hearts again using the stitch of your choice.

Heat Sensitive Stabilizer

Edge finish: sheer ribbon folded and sewn around the edges with a zigzag stitch

1. Use diluted acrylic paints to stain a piece of heat sensitive stabilizer. Let dry.
2. Layer it on top of the desired base fabric. Stitch on the heat sensitive stabilizer and fabric layer with rayon, silk, or cotton thread.
3. Using a low to medium heat setting on your iron, press the heat sensitive fabric. The heat sensitive stabilizer will become brittle and crumble. Brush away the distressed bits. Go slowly with this technique so you don't 'heat away' too much of the stabilizer.
4. Take care with future ironing steps and protect the stabilizer with a heat cloth.

Note: We attached our finished piece to the bottom portion of our Passport page, shown above.

Heat Distressed Fabric Beads

There are many ways to create fabric beads. Heat distressing is one of the easiest. Add stitching as a decorative element to make them more interesting.

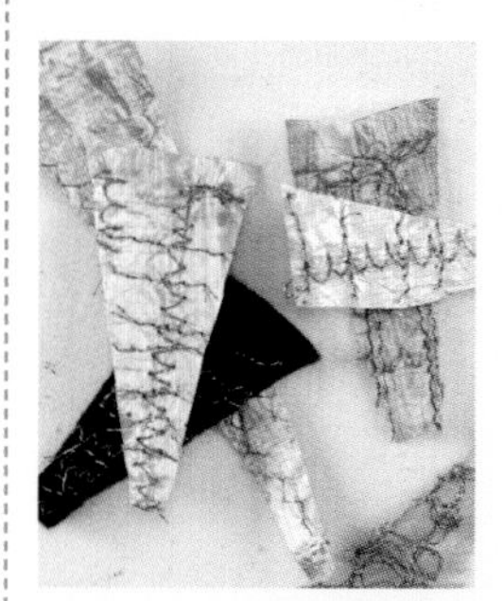

Use scraps of fabric from the exercises on pages 134-136. Add more stitching if needed. Cut the fabric pieces approximately 2" long and 1" wide at the base. Taper to about 1/2".

Start at the widest end and wrap the strip around a skewer. Pin in place. Heat until the bead appears to melt together.

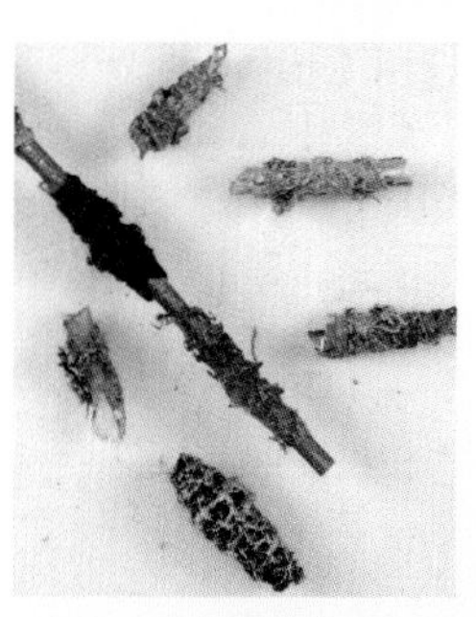

Dry brush the beads with paint to highlight the stitches if desired.

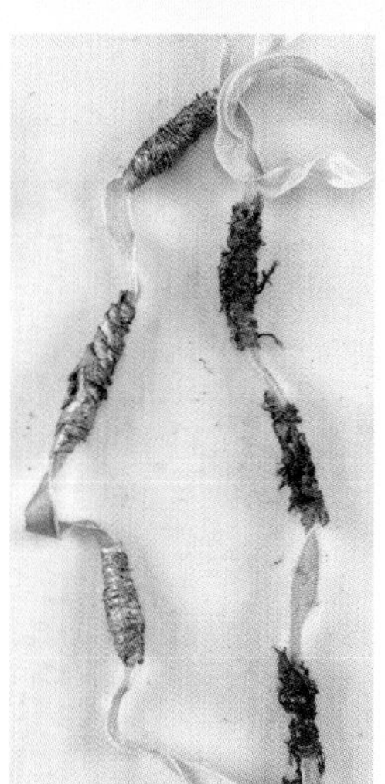

Attach the beads to your passport page or project by pushing a ribbon through the bead's center with a skewer. The ribbon can then be stitched to the page.

machine wrapped cords & sticks

Techniques this easy, versatile, and fun are hard to beat! Machine wrapped cords and sticks use the same method over a wide variety of different materials to achieve ever changing results. These cords and sticks can be color coordinated to any project.

Edge finish: finely wrapped, couched cord, and a zigzag stitch

Edge finish: frayed ribbon folded and sewn on with a zigzag stitch

getting started

Experiment with a variety of fillers such as wooden skewers, heavy threads or strips of fabric twisted together, cord, yarn, or pipe cleaners. The same steps are used for wrapping all the fillers.

Machine set-up:
Drop feed dogs or leave them engaged. Experiment to see which works better for your machine.

Machine foot:
Cording or beading foot

Tension:
Balanced

Threads:
Variety of weights, types, and colors

Needle:
As per thread

1 The stitch covering the filler can be as light or heavy as you like, as long as the filler continues to fit through the channel on the machine foot. Use matching or contrasting thread in the top and bobbin. Medium to heavy weight threads work well. Lower the presser foot onto the filler. Set the zigzag width to clear the filler. Hand turn the hand wheel to be sure the stitch width is sufficient.

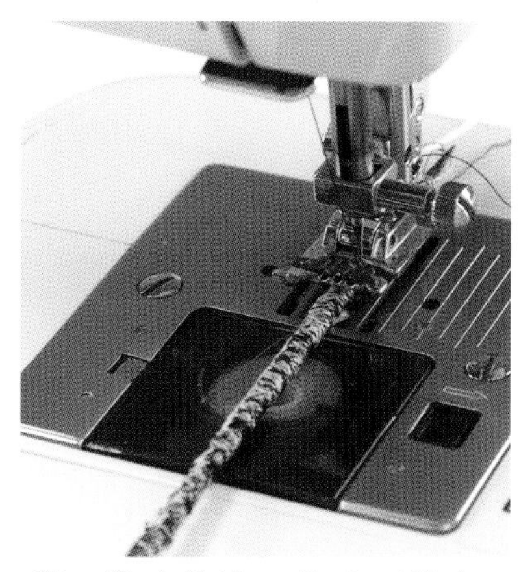

2 Start stitching with about 2" of thread tail at the top of the filler. Steadily move it forward and backward underneath the needle. While moving the filler forward and back, twist it between your fingertips so the top and bobbin threads show on every side. Stitch over the beginning thread tails, securing them within the wrapping. An area of the filler is covered as much as possible when it begins to catch in the foot channel and becomes more difficult to move smoothly. Once this happens along the full length of the filler, it is finished.

3 Use pipe cleaners as a filler to create curves. Follow steps 1 - 2 to machine wrap the pipe cleaners. Bend them into your desired shape and couch them in place on the Passport page.

4 Finish stitching with long thread tails. Knot the tails, snip and rub a dab of glue over the knot. Rub another dab at the other end to prevent unraveling. Attach to your Passport page.

program stitch play

Buttonholes are for more than just making holes for buttons. Make them tiny, make them in pairs, use them for connecting as well as securing.

Edge finish: ribbon couched with zigzag stitch.

getting started

Consult your machine manual for specific directions for creating buttonholes and stitched eyelets. Experiment with different threads for a unique look.

Machine set up:
Normal

Tension:
Balanced

Needle:
Embroidery 90/14 or sharp 80/12

Top thread:
Decorative embroidery or cotton

Bobbin:
Match top thread or use ultra light or light weight thread.

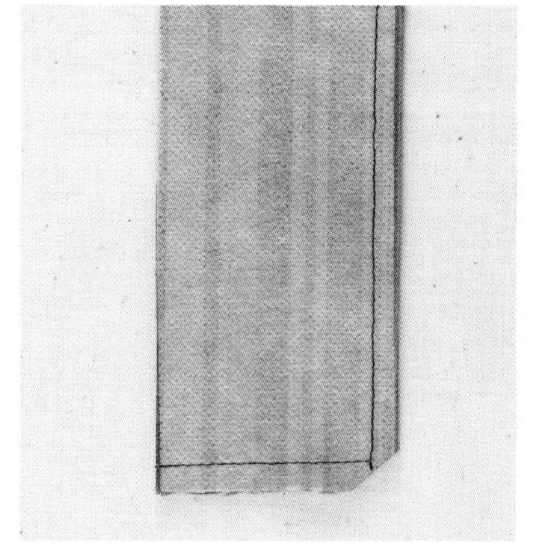

1 BUTTONHOLE LACING: Cut four 2" x 4-1/2" pieces of fabric. If using silk or light weight fabrics use light weight fusible interfacing as a stabilizer. With right sides together stitch around three sides of the fabric leaving one long side open. Clip and trim corners.

2 Turn right side out and press. Make 1/4" buttonholes approximately 1/2" apart along the finished edge. Cut the buttonhole openings. Stitch the open edge closed using a zigzag or straight stitch.

3 Measure the front of the flap to 1-1/4". Press remaining fabric to the back. Place the flaps with the buttonhole edges meeting on the fabric page base and pin in place on the pressed fold line. Check to make sure the front flaps don't overlap. Open the flap out and stitch on the pressed fold line with a straight stitch. Use a zigzag or straight stitch to secure the loose inner edge to the fabric page base. Fold the flaps in and press in place. Lace with cord. A chenille needle is useful for lacing through the small buttonholes.

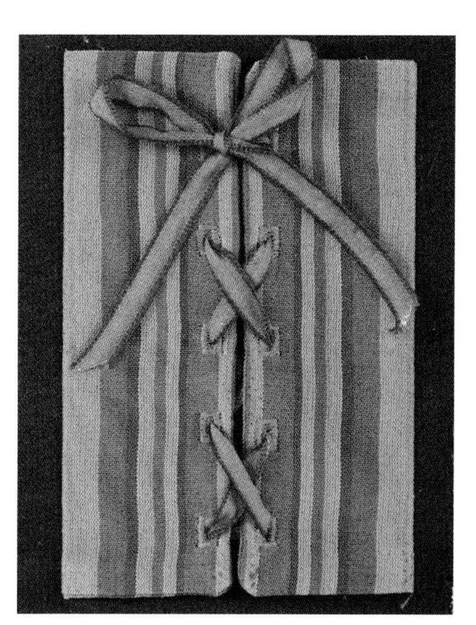

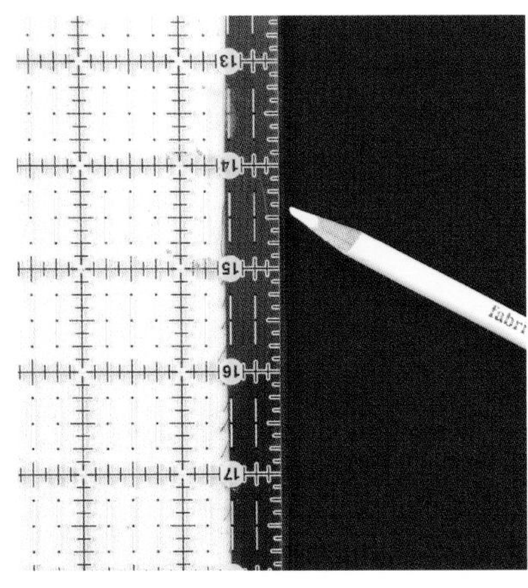

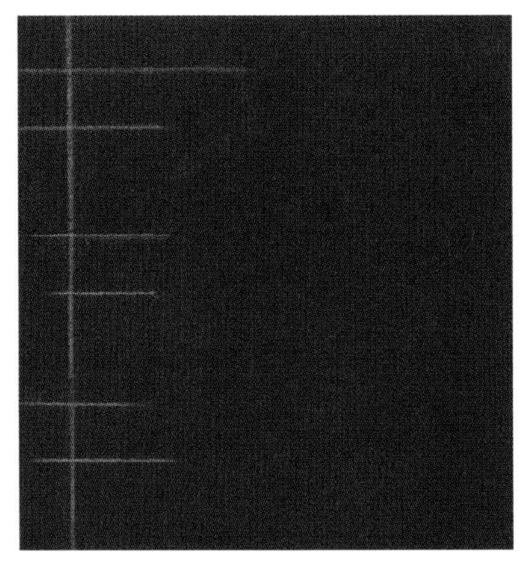

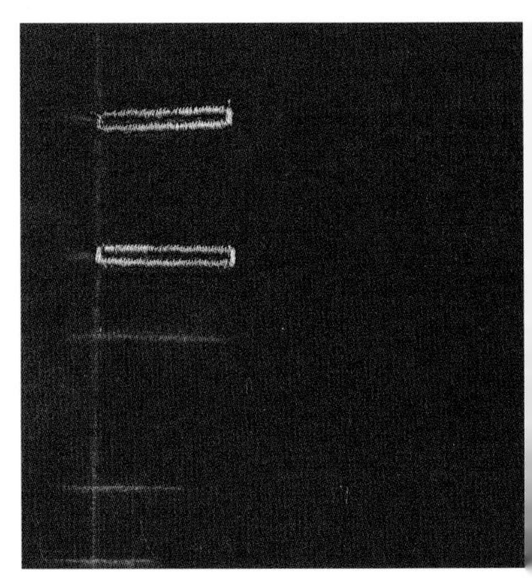

1 BUTTONHOLE WOVEN RIBBON: Choose a ribbon to weave through your fabric page base. Our ribbon is 7/8" wide. To determine buttonhole placement, begin by drawing a chalk or pencil line on the length of your fabric as a guide.

2 Mark buttonhole placement along the chalk line starting approximately 1/2" from the top edge of your fabric page. Mark four pairs of buttonholes 1/2" apart. Leave 1" between each pair of buttonholes. Mark a single buttonhole after the four pairs for a total of nine buttonholes

3 Place a piece of light weight stabilizer on the back of your fabric page and stitch the buttonholes. Cut them open with sharp scissors.

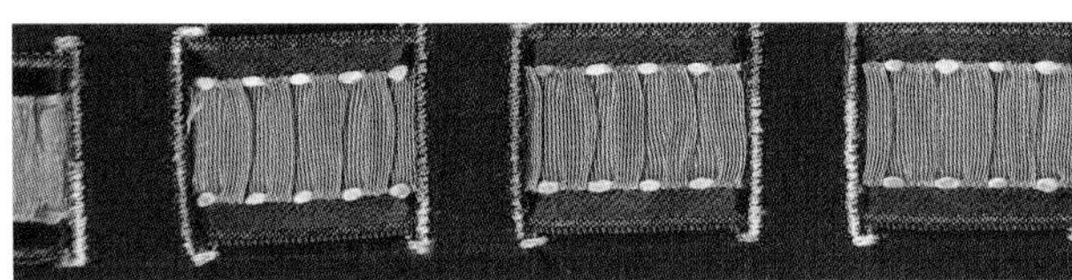

4 Weave your ribbon through the buttonholes. Secure the ends of the ribbon to your fabric with fusible web.

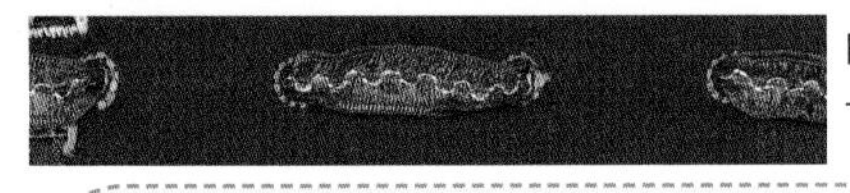

For a slightly different look, weave the ribbon through eyelets following steps 1-4. The horizontal ribbon at the bottom of the Passport page is woven through eyelets.

60-DEGREE SNOWFLAKE GUIDELINE

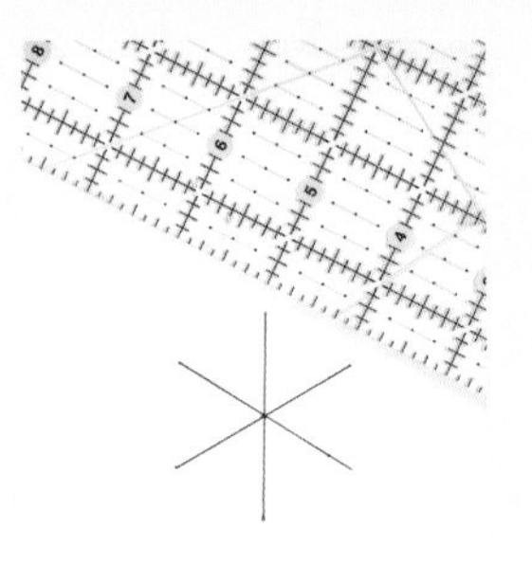

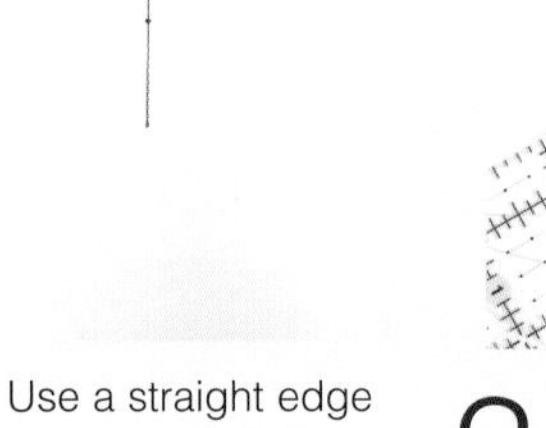

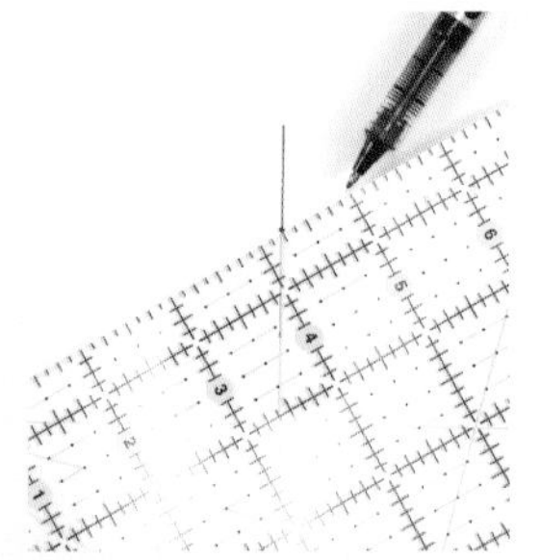

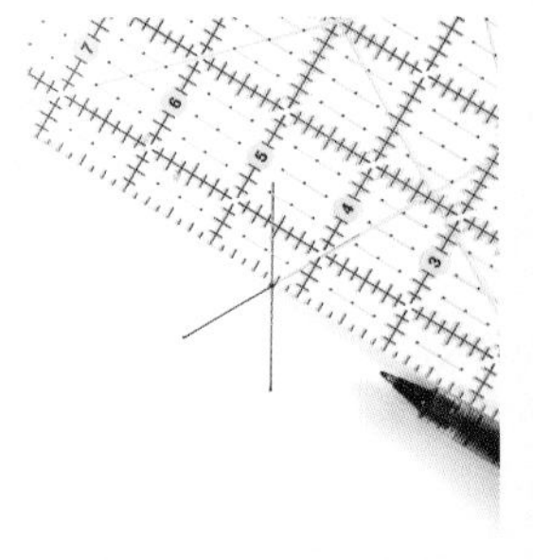

1 In nature snowflakes are created around six-fold symmetry. A ruler with a 60-degree line makes creating a snowflake diagram easy.

2 Use a straight edge to draw a line 2" long. Place a small dot in the center.

3 Place the 60-degree line of your ruler on the drawn line with the edge of the ruler at the center dot. Draw a line approximately 2" long.

4 Move the 60-degree line on the ruler to the new line. Place it at the intersection of the two previous lines. Draw a line approximately 2" long.

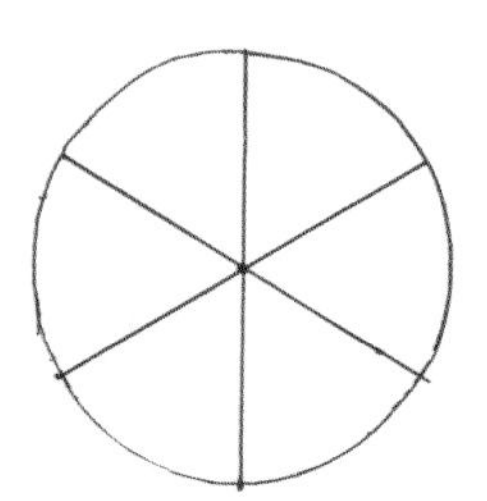

5 To draw a circle perimeter centered over the mid-point of the snowflake, use a 2" circle template. Place the template's guide mark at either edge on the first drawn line that is 2" long. Draw the circumference line on the interior of the guide.

program stitch play

Look at your Programmed Stitch Guide from page 23 for patterns geometric in form - triangles, diamonds, circles, and points. What pattern do they create when stitched back to back in a mirror image or when you alter the length or width? It can take a bit of play and experimentation to find ones that work well together so be sure to make some notes about the ones you like and the adjustments you make for future reference.

Edge finish: zigzag satin stitch in metallic with couched metallic cord on inside edge

1. NOTE: Refer to 60-degree snowflake guidelines on page 142 before beginning. With chalk or a marking pencil draw a 60-degree star guideline on your base fabric. Layer your fabric with a light weight stabilizer.
2. Determine the best method of stitching your chosen design. Does it look better stitched from the outside to the center, from the center out, or can it be stitched across the length of the snowflake?
3. Stitch chosen design.

Machine set up:
Normal

Tension:
Balanced

Needle:
Embroidery 90/14 or sharp 80/12

Top thread:
Decorative embroidery or cotton

Bobbin:
Match top or use ultra light or light weight thread.

FEATHER SNOWFLAKE: This delicate snowflake uses a programmed feather stitch that looks best stitched from the outside to the center. Mark the 60-degree lines. Use a circle template to draw the perimeter circle lines. Begin the stitch at the outer edge of the circle then stitch towards the center. Stitch on all six radiating lines.

POINSETTIA SNOWFLAKE: To create this snowflake the stitch is made by slightly lengthening the stitch on a satin stitch triangle stitch. Stitch the design in one direction for one repeat, pivot 180-degrees and stitch back to the center. Pivot again and repeat along each snowflake guide line. A repeat of a circle satin stitch and a couple straight stitches are added to the tip of each point to finish the snowflake.

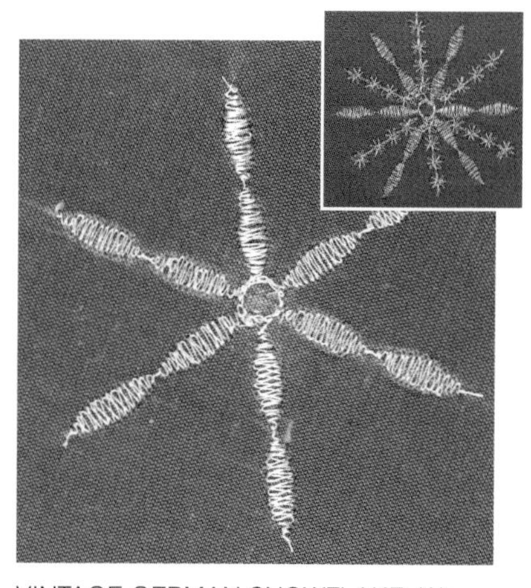

VINTAGE GERMAN SNOWFLAKE: We used a silver metallic thread to give a vintage mercury glass feel to this snowflake. An eyelet stitch was laid down to provide an open center. Then an oval satin stitch at a slightly reduced width was stitched from the edge of the eyelet stitch out for two repeats. A flower chain stitch was inserted between each of the previous stitch lines to create this delicate structure.

program stitch play

Review your Programmed Stitch Guide from page 23 to find stitches that are more organic in feel. Leaves and vines can be combined with a little free motion stitching. Utility stitches can create a delightful landscape. Vines can be stitched in overlapping waves to create a bush.

Edge finish: yarn couched with a zigzag stitch

Machine set up:
Normal and free motion

Tension:
Balanced or slightly loosen top

Needle:
Embroidery 90/14

Top thread:
Decorative embroidery in cotton, rayon, silk, holographic, or metallic

Bobbin:
Ultra fine or light weight

tips

- Transfer your pattern with the aid of transfer marking paper
- Make notes about the adjustments made to your programmed stitches for future reference.
- An open toe embroidery foot is helpful for placing these stitches.
- Back the fabric page base with a medium to heavy weight stabilizer.

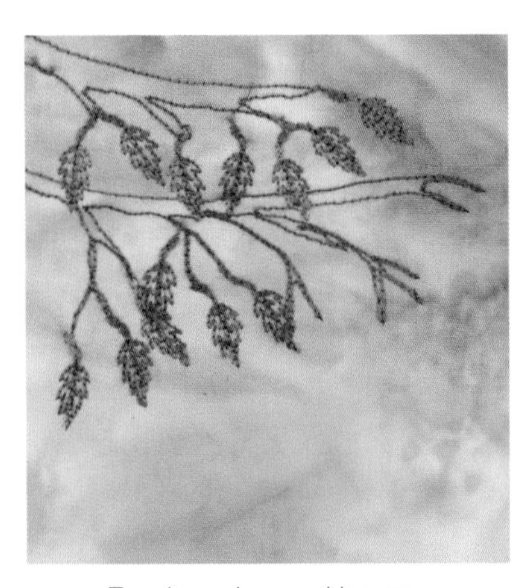

1 Tree branches and leaves: Free motion stitch an outline of a branch on your fabric page base. Design your own branches or use ours as inspiration. Stitch individual leaf motifs hanging from the branches.

2 Vine bush: Stitch two or three lengths of a vine stitch beginning at the bottom of the page and fanning them out slightly as you stitch. We chose a yellow to dark green variegated thread to add some highlights.

3 Stitch over the previous stitching with a solid green thread using the same vine stitch. Stitch from the bottom up, slightly overlapping the previous stitching. Vary the length of the vine to create a natural looking bush.

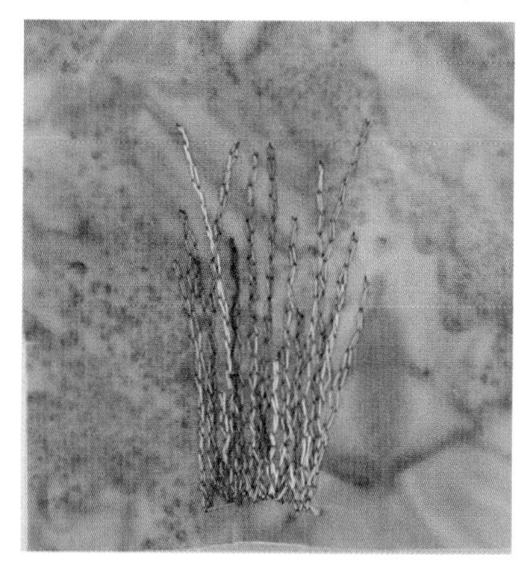

4 Cattails: Change your machine to free motion stitching and select a very narrow zigzag stitch. Stitch the cattails from the base up. Gently curve some of the stitch lines. Using a free motion very narrow zigzag lends a natural rough feel to the cattail stalks.

5 Change your machine to normal stitching. Choose a satin stitch oval shape from your Programmed Stitch Guide. Narrow the width and stitch one repeat of the motif at the tips of some of the stalks. If you don't have a programmed stitch simply begin with a very narrow zigzag stitch set to satin stitch length. Stitch slowly and adjust the stitch width from wider to more narrow as you create the cattail tips.

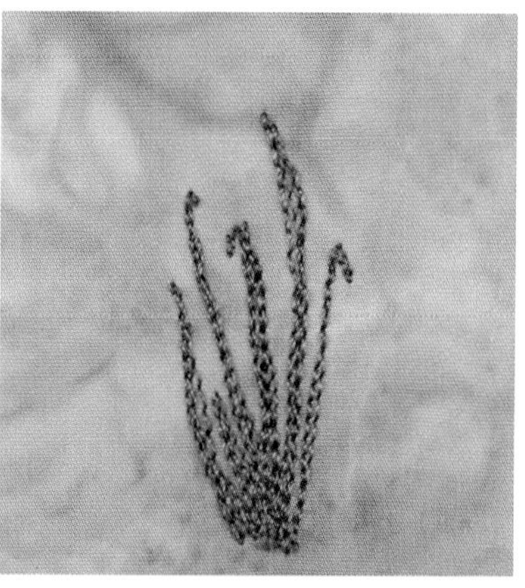

6 Wild Flowers: Set your machine for free motion stitching and use a straight stitch. Begin at the base of the clump of wildflowers and stitch gently curving lines. Vary the height and width of the stems and leaves. The straight stitch will give a smooth thick feel to the leaves of the wildflowers.

7 The tiny flowers are done with a button attachment or tacking stitch. If you do not have this stitch use a zigzag stitch set at a zero length. Use the needle down position and stitch one repeat of the tacking stitch or 8-10 stitches of a satin stitch zigzag. End the stitching on the inside of the flower, pivot the fabric a tiny bit and stitch another set of stitches. Continue pivoting and stitching until you have five or six petals of your flower.

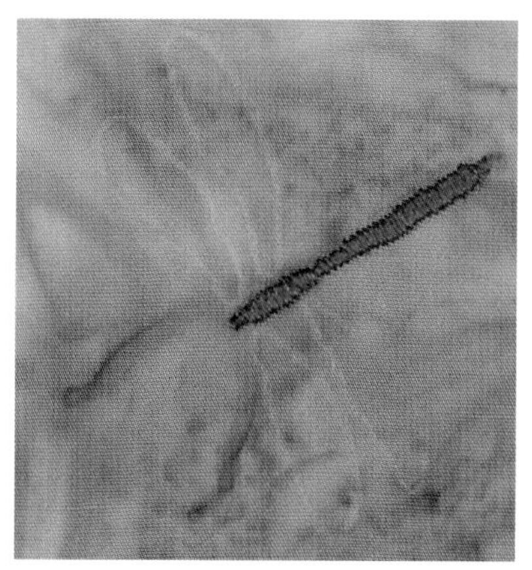

8 Dragonfly: Transfer the dragonfly pattern below onto your fabric or draw your own pattern. Choose a satin stitch zigzag stitch. Begin at one end of the dragonfly body and slowly stitch up the length widening and narrowing the stitch width. A single strand of holographic thread was woven through the satin stitches using a running stitch on the lower half of the body to give a hint of shimmer.

9 The wings are stitched using free motion straight stitching and a reflective metallic thread.

Dragonfly Template

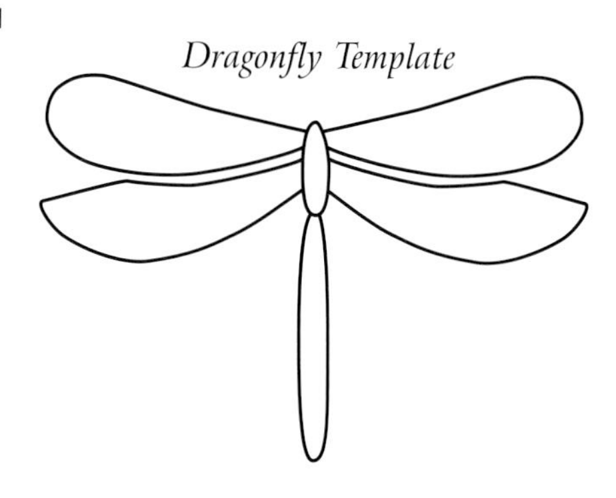

program stitch play

Fringe Grasses: Choose the blindstitch or blind hem foot and a triple stitch set to the widest zigzag and a short length. Try a few samples to see what length gives you the fullest grass but does not build up too many stitches and create a lump. As an alternative you can use a satin stitch zigzag.

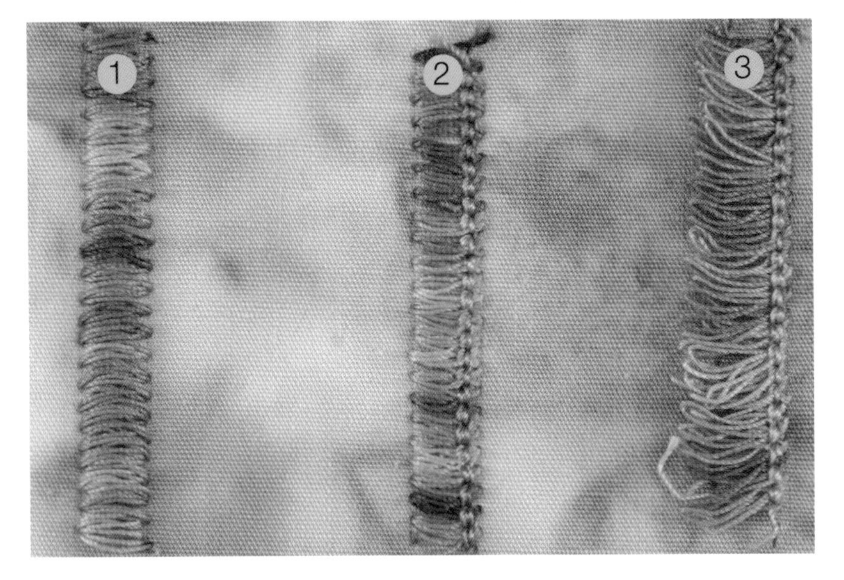

1 Loosen the top tension to zero. Slightly tighten the bobbin tension. Choose a bobbin thread much darker or lighter than the top thread. Insert your blind hem stitch foot. Stitch the desired length using the triple stitch zigzag. Remove the fabric from the machine. The bar on the blind hem foot will result in stitches that are raised above the fabric. The top thread should be pulled to the back. The photo of the back of the fabric shows very little black bobbin thread.

2 Change to a straight stitch foot and either a triple straight stitch or shortened straight stitch. Stitch along the bottom edge of the previous stitching. This stitch will secure the stitches.

3 Turn the fabric over and clip away the bobbin threads. From the right side of the fabric, pull the loose top threads to the front and fluff.

tip

- To stitch multiple rows of grass close together begin with the row of grass that will be in the foreground. After that row is completely finished (steps 1 - 3), press the stitching away from the next row. Secure with tape if needed. Stitch the next row close to the first.

Add fringe grass to your Passport page.

Box of Secrets

Interlocking halves of a heart form the closure to this box which is confetti stitched over torn strips of fabric. A combination of free-motion and decorative stitching covers the fabrics. Machine wrapped, bead embellished pipe cleaners are tacked to each corner.

Confetti Fabric Strips

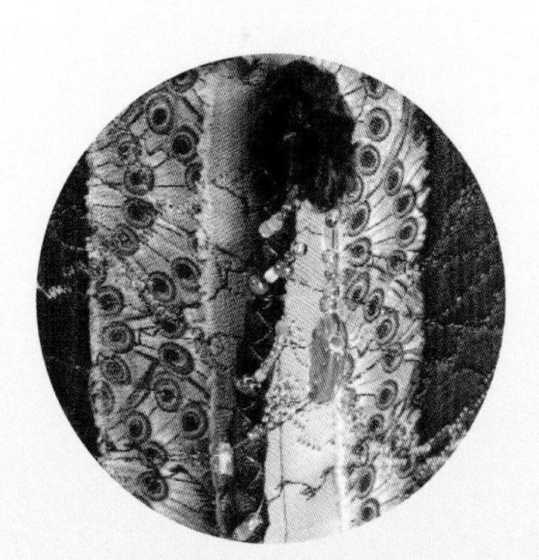

Machine Wrapped Pipe Cleaners

Beading

sashiko inspired stitching

Sashiko, or "small stabs," is the traditional form of Japanese stitching. Although the designs are often ornate and intricate, the stitch is a variation of the humble running stitch. These stitches are not traditional Sashiko but are inspired by the use of heavy threads in straight stitching with a Japanese flair.

Edge finish: a strip of antique kimono silk is stitched around the edge

getting started

Sashiko stitching usually includes sharply defined designs and plump, evenly spaced stitches. This piece is hand stitched and then machine stitched to the base with the hand look quilting stitch. It is finished with a Japanese-style crest. We challenge you to design your own creative mark.

Machine set up:
Normal

Tension:
Balanced

Needle:
Hand 8 and
Embroidery 90/14

Top thread:
Hand - heavy weight;
Machine – monofilament

Bobbin:
Medium weight

1 Trace the kimono template onto freezer paper and cut out. Iron the pattern to the base fabric, shiny side down. Using a chalk marker, trace the outline of the kimono and remove the paper. Apply fusible web to the back of a decorative fabric scrap. Trace a spool end or other small circle onto the paper backing.

2 Cut out and fuse the fabric circle within the kimono outline. Hand stitch the outline of the kimono along the chalk line. Stitch a second line along the hemline as a decorative detail. Try to achieve a consistent stitch size. Do not pull stitches tight.

3 Trim the square to fit the page as desired. Fray the fabric if you like. Attach the kimono square to the Passport page base referring to the Hand Look Quilt Stitch on page 62.

Add Your Own Touch

- Design your own *Mon*, or family crest, on the corner of your Passport page.

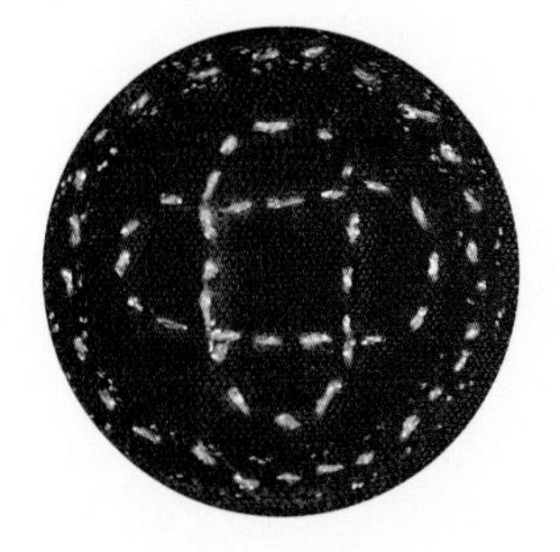

Did you know?

- In the Hari-Kuyo ceremony, Japanese women gather once a year to thank their worn out needles and pins for good service. It is also a time to value the small, everyday objects of daily living and to wish for progress in one's needlework. In what is sometimes known as the Festival of Broken Needles, women gather to offer a funeral-type service by laying the needles to rest in soft jelly cakes or tofu. This burial is meant to bring rest to the needles and wrap them with tenderness and gratitude.

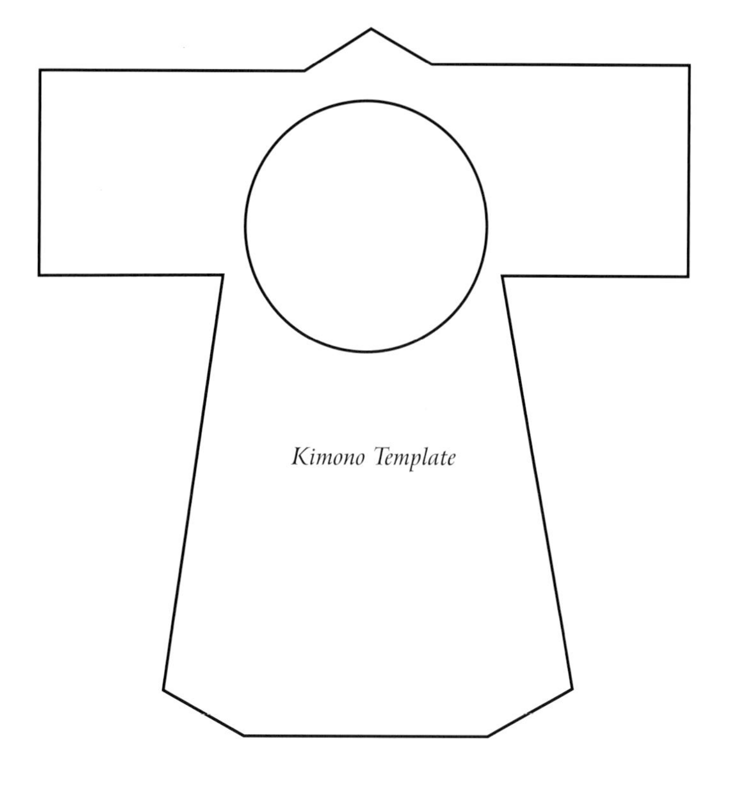

ribbonry

There is something about ribbon that just makes you want to use it!
Ribbons add a splash of color and can hang freely and move.
They add a touch of whimsy and are perfect for stitching on.

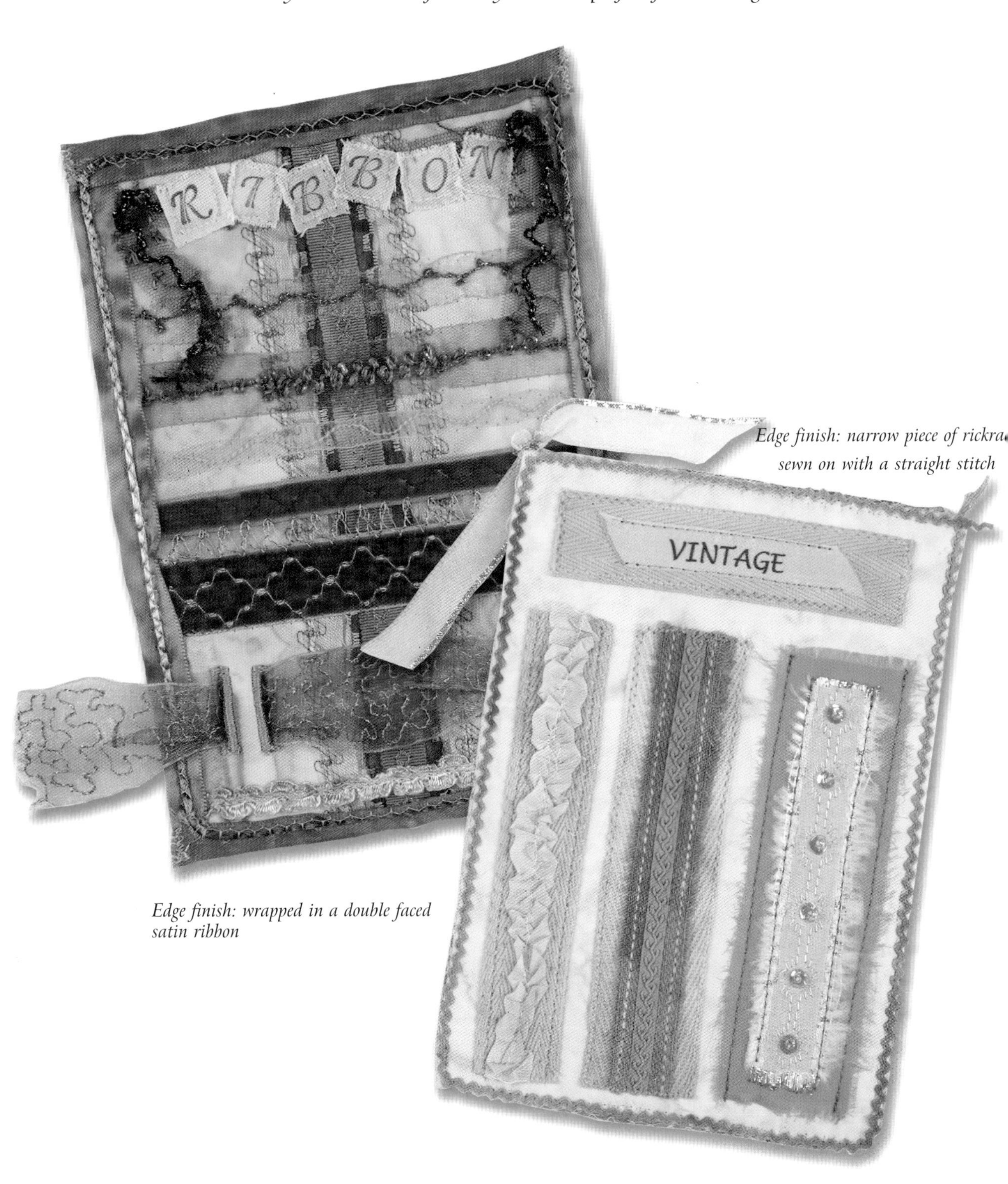

Edge finish: narrow piece of rickrack sewn on with a straight stitch

Edge finish: wrapped in a double faced satin ribbon

getting started

By layering one stitched ribbon atop another, even the simplest of ribbons can become a fancy trim.

Machine set up:
Normal

Tension:
Balanced

Needle:
Embroidery or sharp sized to the thread. Optional, a twin or triple needle for decorative sewing

Top thread:
Medium and heavy weight decorative

Bobbin:
Fine or medium weight

Make Your Own Trim: Place a medium width ribbon on larger width ribbon and edge stitch both sides with a decorative stitch. Center the narrow ribbon on the previously layered ribbons. Stitch in place with a decorative stitch. Refer to your machine's manual to set your machine up for triple needle stitching. Stitch your ribbon to the base fabric.

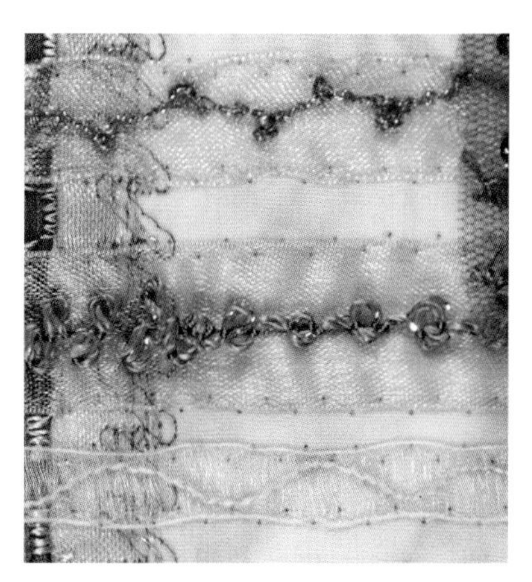

Create Texture on Organza Ribbons: Stitch organza with a variety of heavy threads. Refer to Sewing with Ultra Heavy Threads on page 58.

Machine set up:
Free motion and buttonhole

Tension:
Slightly loosen upper

Needle:
Embroidery 90/14

Top thread:
Medium or heavy weight decorative thread.

Bobbin:
Medium weight

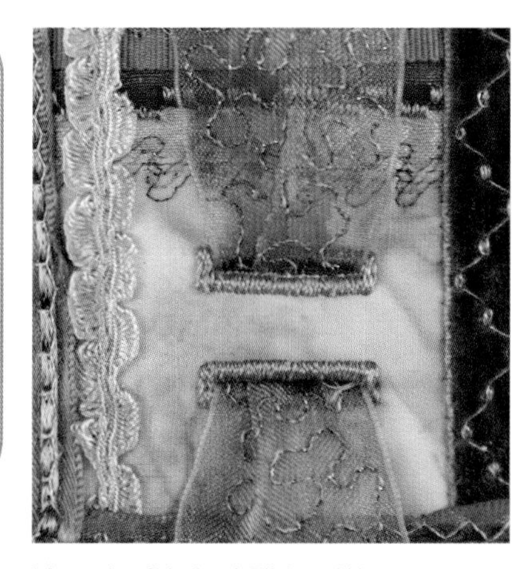

Meander Stitched Ribbon Woven Through Buttonholes: Hoop 8" of ribbon so it is taut and sew with a decorative or stipple stitch. Refer to your sewing machine manual to make two buttonholes on the base page. Weave the ribbon through the buttonholes.

Decorative Ribbon: Use a simple straight stitch in monofilament or ultra-fine thread to add decorative ribbon to your page. When stitched with a lengthened straight stitch down the center the textured curls will not be flattened.

ribbonry

Machine set up:
Normal

Tension:
Slightly loosen upper tension

Needle:
Embroidery 90/14 or what works with chosen thread

Top thread:
Medium or heavy weight decorative

Bobbin:
Medium weight in a contrasting or blending color.
The bobbin thread will be visible.

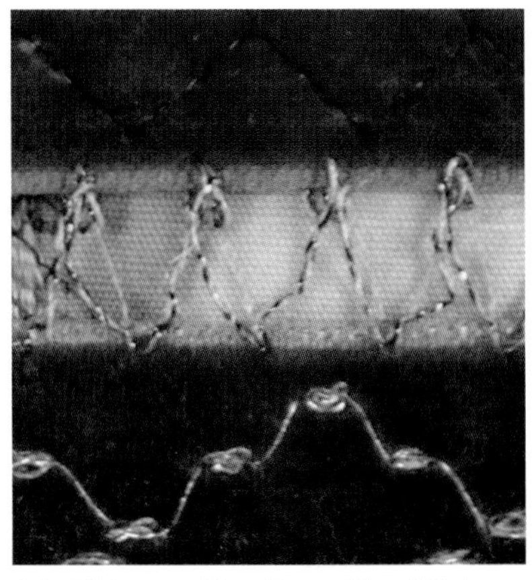

Join Ribbons with a Faggotting Stitch: This is a great "special effects" technique. Most sewing machine companies offer a faggotting foot, possibly called a fringe or tailor tack foot. A blind hem stitch foot can also be used. Zigzag stitches in two different metallic threads are held up on a ridge between two ribbons which are then tugged apart to flatten the stitches. If you do not have this foot, try using a coffee stirrer to keep a narrow gap between two ribbons while zigzagging them together. Abut two ribbons, one on either side of the sewing foot, and stitch with a wide or medium length zigzag or triple zigzag stitch. Sew slowly down the length of the ribbons. Make sure the zigzag takes a small bite into each ribbon edge. End with a locking stitch. Gently tug the two ribbons to each side to flatten the joining stitches. Attach the ribbons to your Passport page base.

Stitching Velvet Ribbon with Metallic Threads: Add metallic stitching to velvet ribbon and it becomes lush.

Torn Fabric Strips as Ribbon: Our ribbon label was made by bobbin stitching ultra heavy thread onto a torn strip of purple lace. The letters are rubber stamped on fabric and then stitched over the lace on the background page. The tails are allowed to hang freely. Refer to Rubber Stamping on page 170.

Edge finish: wrapped with bias cut silk ribbon

1 A torn piece of silk is laid on the background fabric with a narrow ribbon on top. Both layers are stitched to the background fabric with a wavy stitch.

2 Use thread painting to mimic a design from an ornate woven ribbon or fabric. Refer to Thread Painting on page 66.

3 Narrow silk ribbon can be used like ultra heavy threads through the bobbin. Refer to Sewing with Ultra Heavy Threads on page 58.

4 The twin needle is perfect for tacking a ribbon in place. Top it with a bobbin-stitched twisted ribbon for embellishment. Refer to your sewing machine manual for information on threading the twin needle. Twist the two-sided ribbon and hand tack in place.

5 Gathering works best with a fine ribbon like organza. Use three times the final length desired and a longer stitch length. Lock your beginning stitch, sew the length of the ribbon, hold the ending thread and gather. Knot the end thread tails and clip.

tips

- Check the affect of heat on ribbon. Many ribbons melt at low iron temperatures.
- Use clear nail polish to seal ribbon ends.
- Use fusible web tape to fuse layers of ribbon before stitching to prevent shifting.

tassels & fringe

Tassels don't require extra equipment. All you need is some cardboard and thread.

Edge finish: knobby yarn couched with a zigzag stitch

We have shown the same tassel with 5 simple finishes.

1. Neck wrapped with cotton yarn and one strand of the same yarn for the hanger.
2. Neck wrapped with contrasting wool/acrylic thread and multiple strands of the same thread for the hanger.
3. Short lengths of tape yarn were glued at the neck to form a skirt over the tassel. Black perle cotton was wrapped around the neck and used for the hanger. Two wraps of a contrasting thread were added around the neck.
4. Bead fringe was added at the base of the neck. Work from the inside of the tassel out through the bottom edge of the wrapped neck. The neck was wrapped with a knobby yarn and the same yarn used for the hanger.
5. Before removing the thread wraps from the cardboard a second layer of wrapping in an analogous color was added to create a skirt. The neck and hanger are silk yarn.

Edge finish: organza ribbon couched with a zigzag stitch

Tassels may be made with any type of thread or yarns. We have shown 6 different examples.

1. wool/acrylic thread - medium weight
2. metallic/rayon thread - heavy weight
3. variegated cotton thread - medium weight
4. cotton thread - heavy weight
5. 2 colors of silk thread - medium weight
6. rayon thread - heavy weight

getting started

Tassels and fringe don't rely on strong thread so this is a great project to use up some older thread that has lost its tensile strength over time.

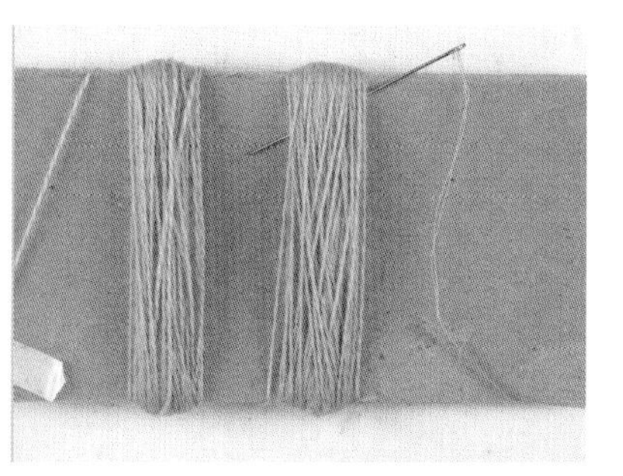

1 Cut a piece of cardboard the length of your desired tassel and between 4" and 6" wide. The width is not important unless you want to make multiple tassels at one time. If your desired tassel length is 2" cut a piece of fairly stiff cardboard 2" by 4". Tape or hold one end of the thread out of the way along the bottom or side of the cardboard. Wrap the thread around the cardboard overlapping the threads in a space about 1/2" wide. The number of wraps needed depends on the thickness of the thread. The wool/acrylic thread used here required approximately 125 wraps. Finer thread will require more wrapping. When making multiple tassels using the same fiber simply move to the right about an inch and begin wrapping again. When the wrapping is finished, thread a needle with a short length of thread (about 6") and slide the needle between the wrapped thread and the cardboard. Remove the needle and tie the thread in a square knot around the thread bundle.

2 Cut the tassel thread between the wrapped sections and remove the thread bundle from the cardboard. You may have to slightly bend the cardboard to slide the thread off. Cut a 30" piece of strong cotton thread. Grasp the thread bundle about 1/4" - 3/8" from the top. Lay the cotton thread along the length of the thread bundle with about 1-1/2" free of the top of the bundle.

3 Grasp the top of the bundle in one hand and begin to wrap the long length of the cotton thread back up the neck of the tassel towards the top. Wrap very tightly. Continue to keep the short length of thread free as you wrap the thread back over itself. The neck will be about 1/4" long.

4 When you have wrapped up to the top of the neck, use a needle to take the two loose ends of wrapping thread under the wrapped neck and down toward the bottom of the tassel. Pull these threads tight to secure the tassel neck. Clip off extra thread.

6 Use scissors to cut the bottom loops of thread. Fluff your tassel and even up the bottom with scissors if needed. Thread a large eye needle with a hanging thread or yarn and pass it through the top knot following the path of the original thread holding the bundle together. After you have added your hanging threads you can cut the thread that tied the bundle together and pull it out.

tips

- Spray the threads with a fine mist of water to settle the fibers into place.
- Use a small amount of tacky glue around the neck to secure.
- Variegated threads make wonderful tassels.
- Use strong cotton thread to wrap the neck. Slippery threads will not hold.

tassels & fringe

SIMPLE STITCH FRINGE is fast to create and has a unique funky look. Make it with water-soluble stabilizer or a fringe frame. Set up your machine as for Thread Lace Grid on page 110. Refer to that section for tips and basic steps.

Edge finish: ribbon couched with a zigzag stitch.

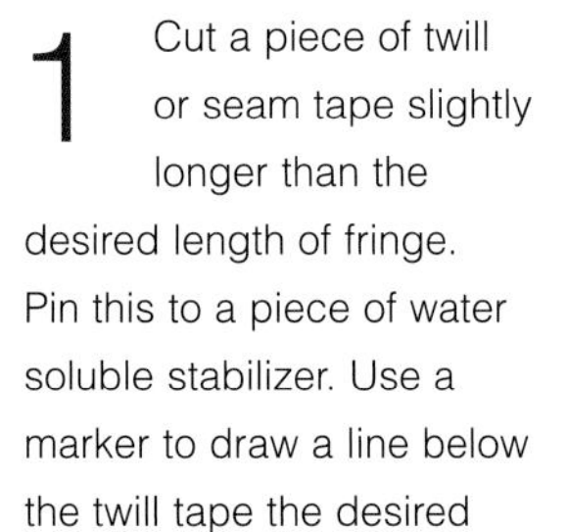

1 Cut a piece of twill or seam tape slightly longer than the desired length of fringe. Pin this to a piece of water soluble stabilizer. Use a marker to draw a line below the twill tape the desired width of the fringe. Stitch the twill tape to the stabilizer and remove pins. Free motion stitch straight lines from the twill tape down to the marked fringe width line. Stitch as closely as possible. Try not to overlap lines of stitching. Rinse out the water soluble stabilizer. The fringe can be cut or left loopy. Refer to the top two samples on the Passport page.

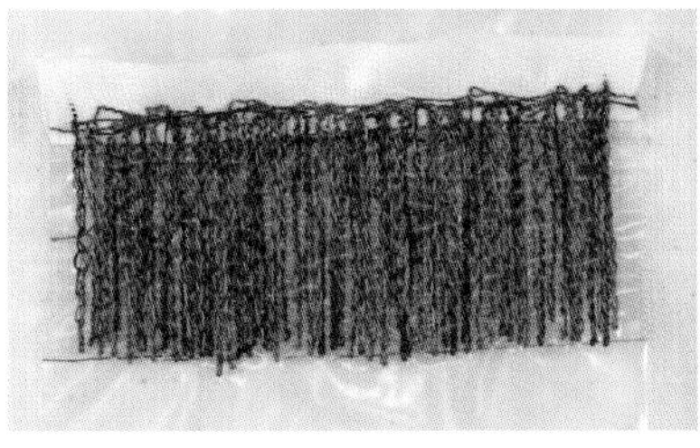

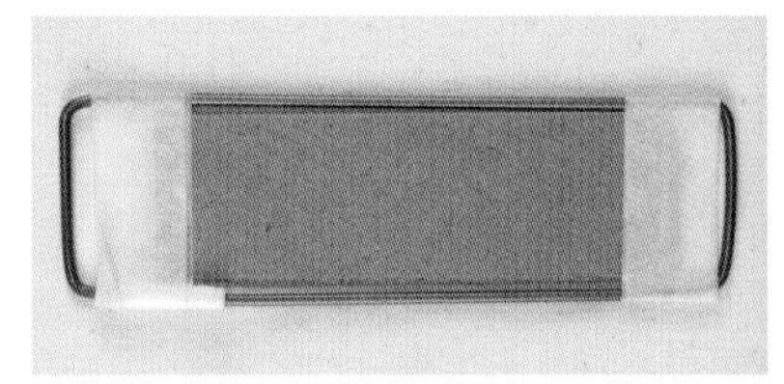

2 To make a fringe frame use wire cutters to cut apart a wire hanger. Determine the size frame needed to create the desired fringe length and width. Our frame began as a 13" piece of wire. Using pliers, bend the wire 1" from the end and down 90 degrees to form the side. At 1-1/2" bend the wire 90 degrees to form the bottom edge. Measure 5" and bend the wire up 90 degrees. Measure 1-1/2" and bend the wire 90 degrees again. The wire should form a rectangle and the ends should just barely meet. Wrap the opening with masking tape. Cut a piece of cardboard slightly smaller than the inside of the frame in height and about an inch shorter. Tape the cardboard inside the frame.

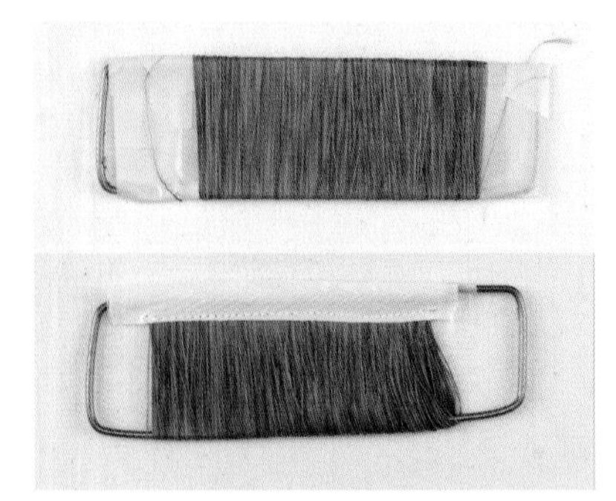

3 Secure the thread at one end of the frame with a piece of tape and begin wrapping the frame. The threads should be adjacent to each other rather than overlapping. Wrap the threads closely but not too tightly on the frame. Continue wrapping the thread until the fringe is the desired length. Use a piece of 1/4" fusible tape and fuse a piece of twill tape to the top edge of the wrapped thread. Remove the masking tape and carefully slide the piece of cardboard out of the frame.

4 Use the zipper foot to stitch along the edge of the twill tape. Stitch multiple times to secure the thread loops. Remove the masking tape from the wire and slide the fringe off the frame.
Stitch again at the top edge of the fringe. The bottom edge of the fringe can be cut or left in loops. To create your Passport page wrap the top edge of the fringe with fabric and stitch or fuse in place.

Refer to the bottom two samples on the Passport page.

tassels & fringe

FANCY FRINGE can be added to the edge of your fabric and is a great way to finish a scarf or shawl.

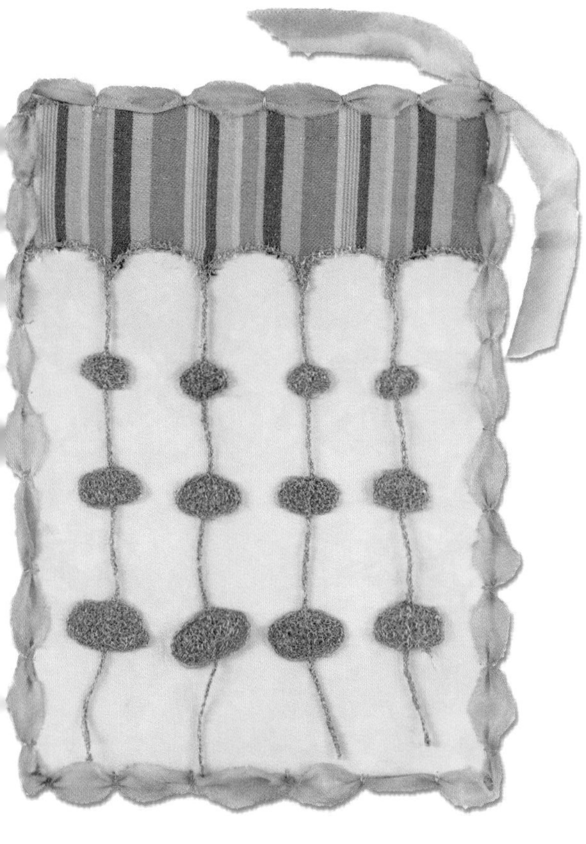

Edge finish: silk ribbon couched by hand.

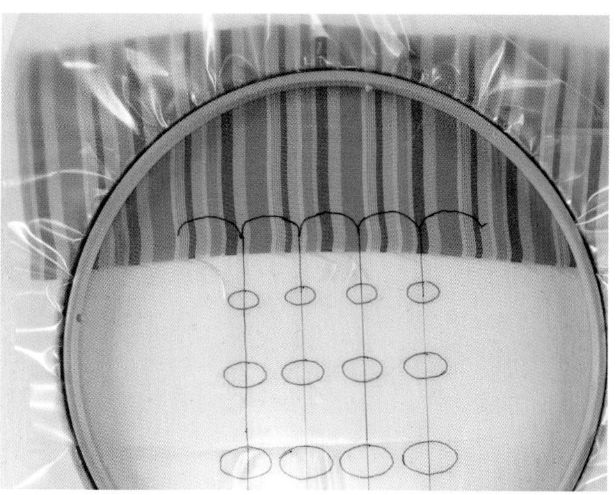

1 Cut a piece of heavy weight water-soluble stabilizer large enough to fit in a hoop. Draw the desired pattern on the water-soluble stabilizer. Cut a piece of fabric approximately 4" x 10". Lay this fabric under the stabilizer and hoop the two pieces.

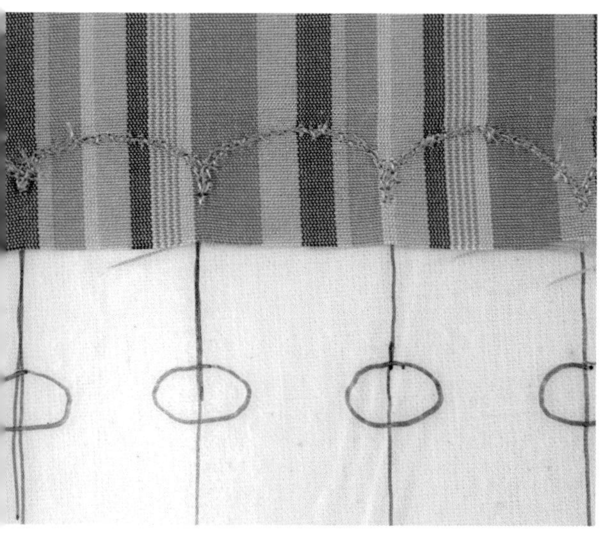

2 Using a free motion straight stitch, stitch along the drawn line marking the edge of the fabric. Make small stitches. Go back over the stitch line with a narrow and short zigzag stitch.

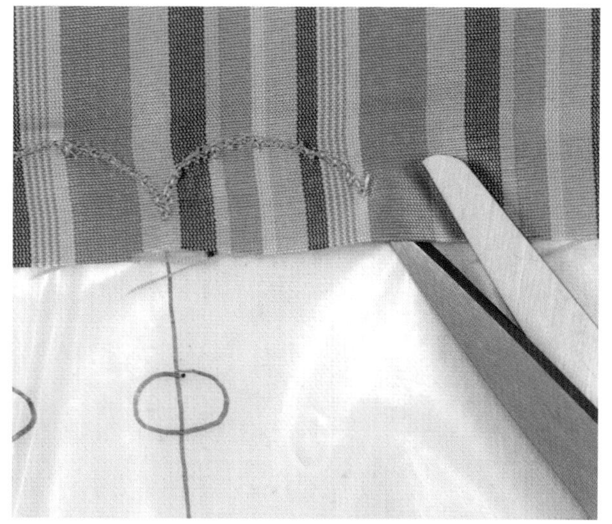

3 Use scissors to cut out the excess fabric. Do not cut away the stabilizer. Cut as close as possible to the fabric without cutting into the stitches. To stitch the fringe, follow the directions for Thread Lace Grid on page 110. Be sure to zigzag over the long stitch lines to bundle the threads and create a base grid in the ovals before filling them in. Rinse out stabilizer and let dry.

the techniques

beading

scraps & metal

adding text

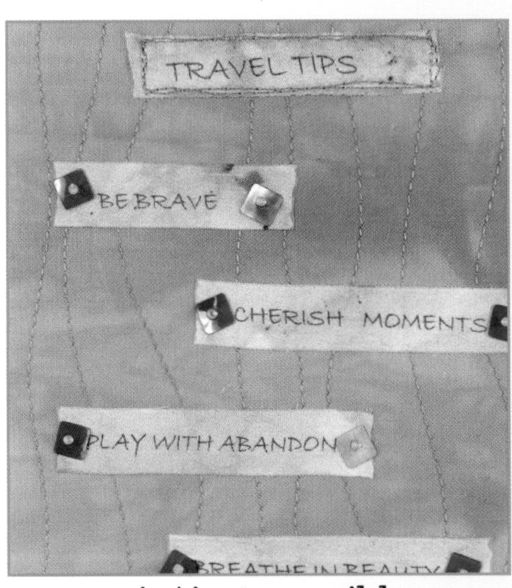

printing on ribbon

adding pockets, stamps & tags

crazy quilting

Thread as Embellishment

Think flourish. Think pizzazz. Adding little touches to your Passport is when you really get to put your personal mark on your work. Take time to add your style to the pages you have created. Whether it is with beading, stamping, or alternative materials, let your originality shine.

beading

Adding beads is a quick and fun way to embellish your Passport pages. Even a few beads scattered across a page can have a dramatic impact.

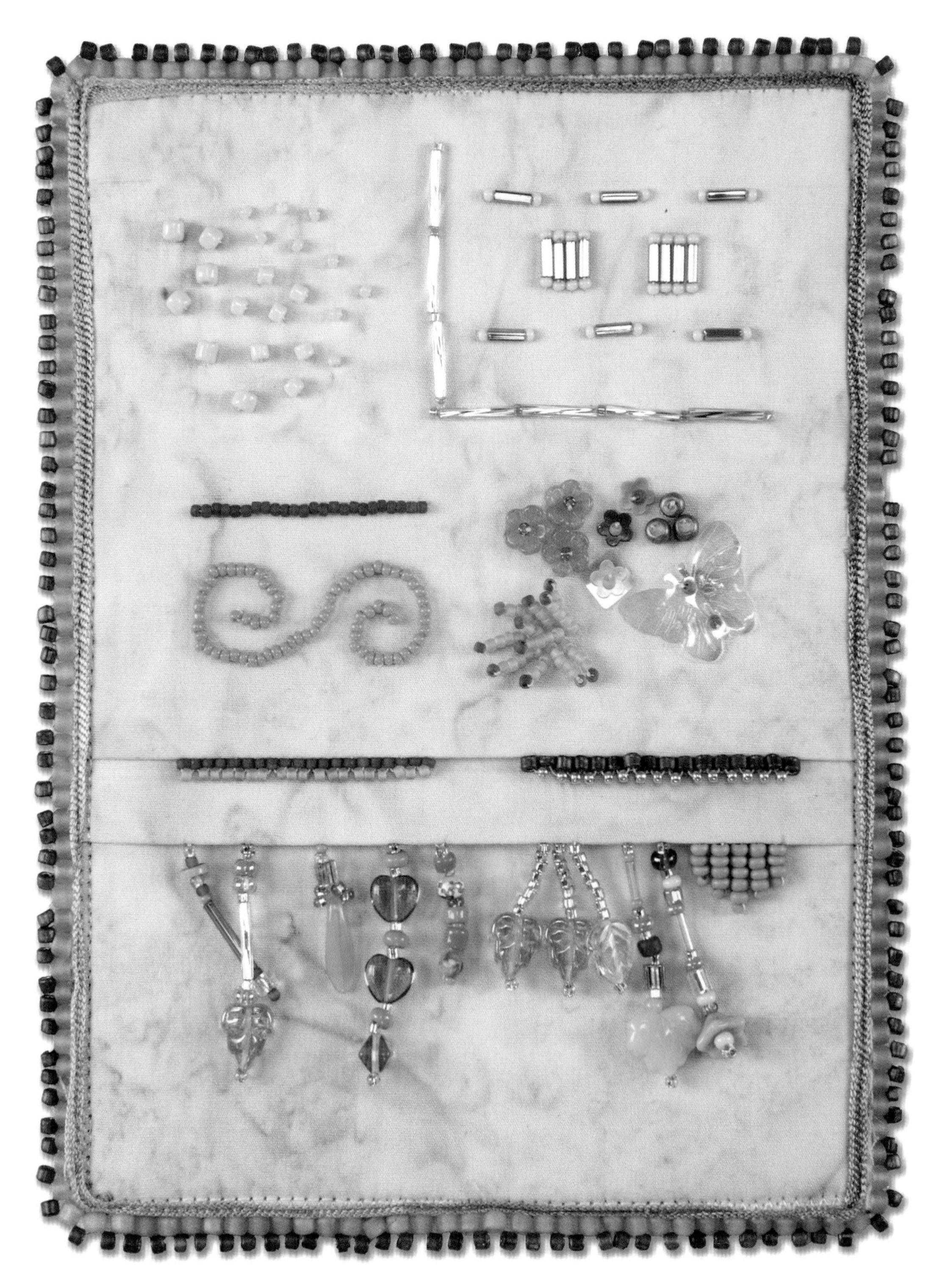

Edge finish: rayon tape yarn stitched on edge with a straight stitch and picot edge added.

getting started

Prepare your page base by cutting a 5" x 9" piece of fabric. Make two tucks or folds on the bottom half of the fabric. The bottom tuck should begin about 2-1/2" up from the bottom of the fabric and the next tuck 1/2" above the first.

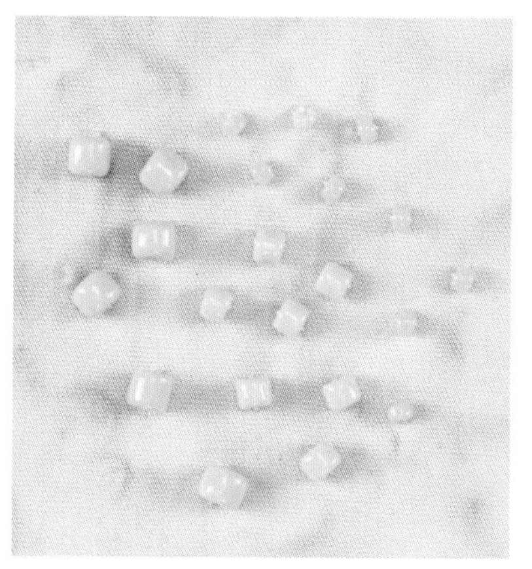

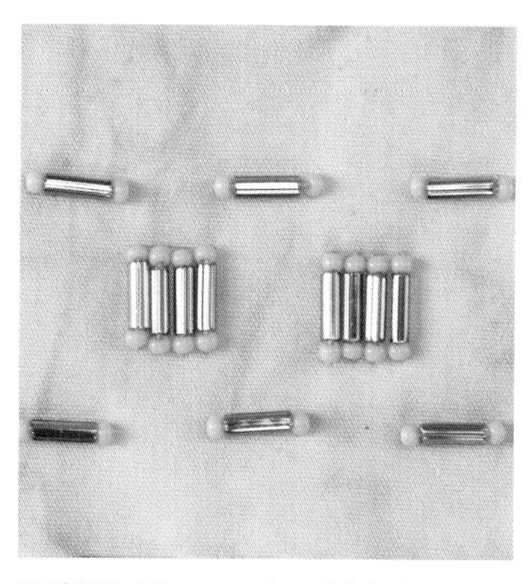

SEED OR SCATTER STITCH: Bring your needle up from the back of the fabric. Thread on one bead and let it drop to the fabric base. Use the side of your needle to push the bead snug against the entry point of the thread. Put your needle straight down into the fabric, pull the thread firmly through but not too tight. Taking the needle straight down and not at an angle will keep your beads sitting up tall instead of flopping to the side.

BUGLES: When used on fabric bugle beads should be stitched with a seed bead at each end. The sharp edges of the bugle beads will cut through the thread over time. Bring your needle up from the back of the fabric. Thread on one seed bead, one bugle bead, and one seed bead. Snug up as with the seed stitch and put the needle straight down into the fabric and out the back. Bugle beads can be stitched in lines, groups, and curves.

BACKSTITCH: The backstitch is useful for laying down straight or curving lines of beads. Beads may also be couched down but this is generally not as secure as the backstitch. Bring your needle up from the back of the fabric and load four beads on it. Snug up as with the seed stitch and take your needle straight down through to the back of the fabric. Bring the needle up between the second and third bead, thread it through the last two beads (three and four). Add four more beads and repeat.

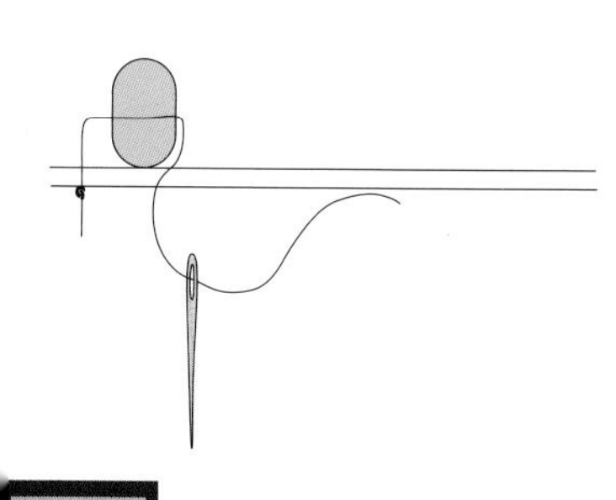

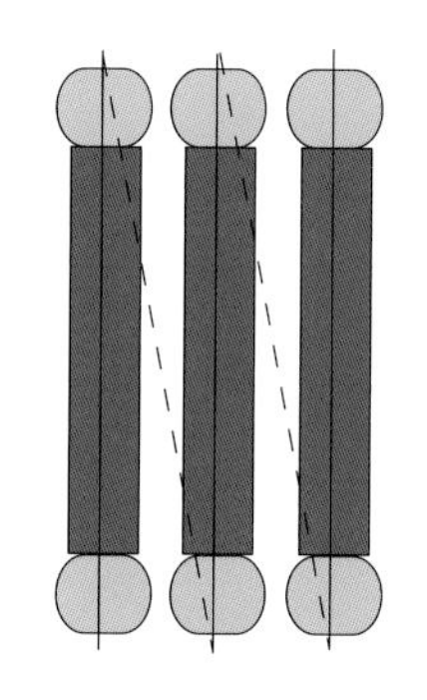

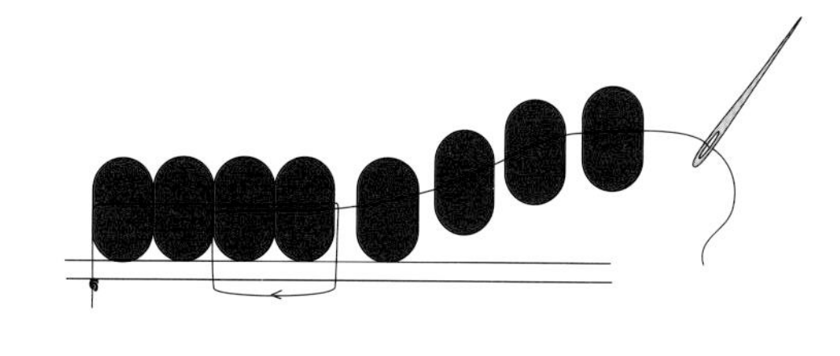

tip

- Beading thread differs from sewing thread. Depending on the beading task or sharp edges on the beads, you might need a stronger thread. Beading thread is sized alphabetically, A – G, finest to heaviest. The two most commonly available beading threads are Nymo and Silamide. Nymo comes in multiple weights and is a flat nylon thread. Silamide is generally found in one weight and is a round thread.

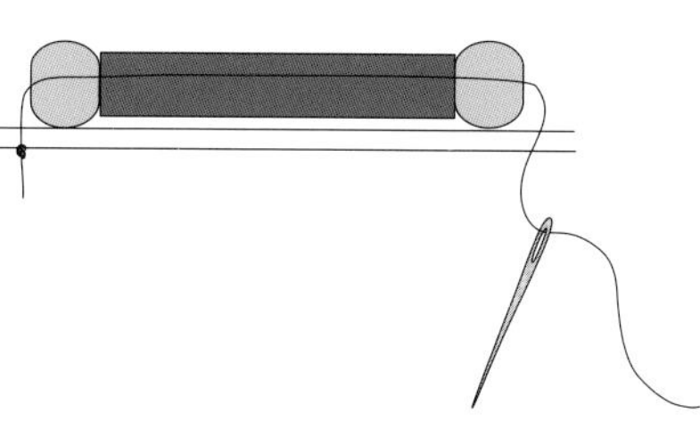

tips

- If your beading will be subject to wear and abrasion tie a security knot every few stitches.
- Cut beading thread with very sharp scissors and at an angle to foster easier threading.

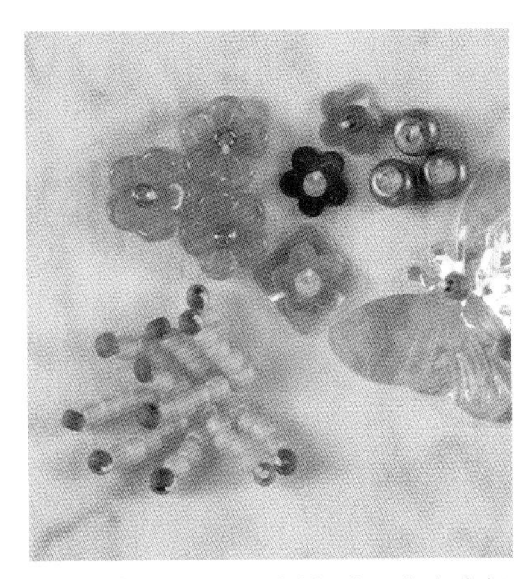

STACKS: Stacks are fairly short in height and are a great way to add sequins, flower beads, and other interesting beads to your work. Stack up to five or six seed beads for unique textures. Bring the needle up from the back of the fabric. Add on the beads you want to stack. Add on one seed bead as your 'stop' bead. Hold on to the seed bead with your non-dominant hand and take the needle back through the remaining beads to the back of the fabric.

PICOT EDGE: This edge finish is fun and adds texture to the edges of your work. Bring the needle out of the edge of the fabric. String three beads. The first and third bead should be the same size. The middle bead which becomes the 'up' bead may be the same size or smaller. Insert the needle one bead wall width away from the base bead. Take the needle back through the folded edge. Bring the needle back up through the last bead and add two beads this time (a middle 'up' bead and a bottom bead). Repeat.

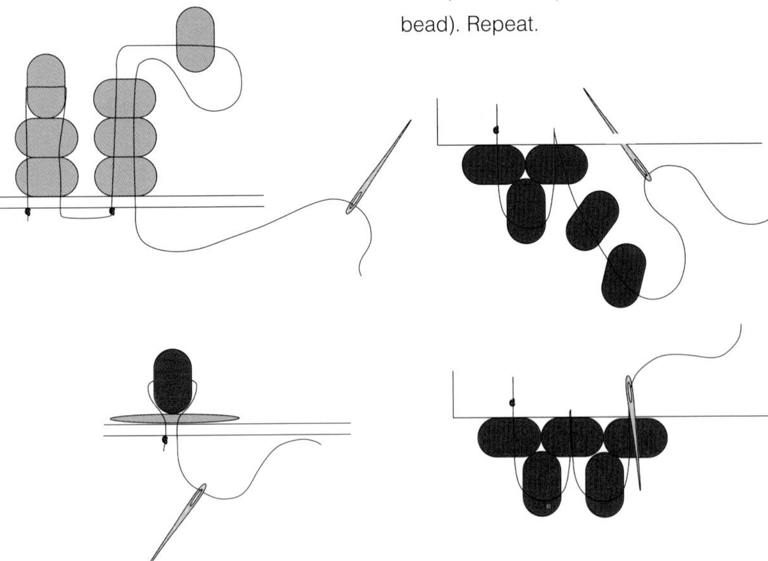

FRINGE: Bead fringe can be worked in so many ways and with such a variety of beads that it can become an obsession. Fringe is a longer stack stitch made at the edge of the fabric. Bring the needle out at the edge or fold of the fabric, add beads as desired. Add a 'stop' bead and hold it in your non-dominant hand. Take the needle back through the rest of the beads and the back of the fabric.

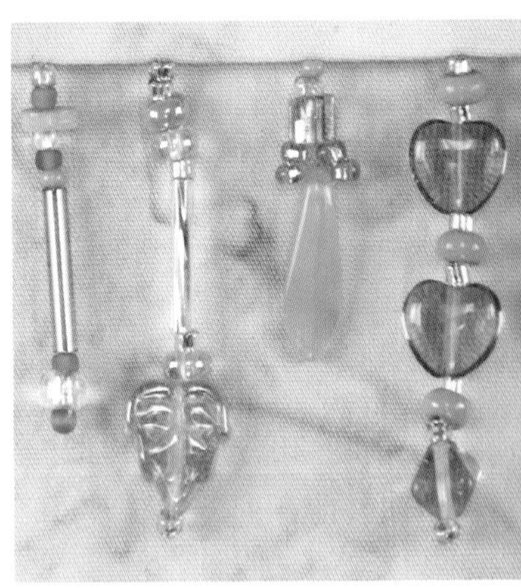

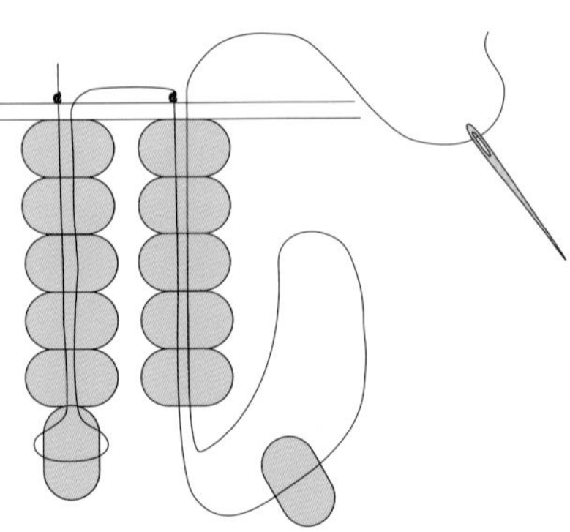

Danse

Working in small bits and then joining them together give you more freedom to create without worrying. In Danse, small bits with rubber stamped trapunto and free motion thread writing are completed and then added to the full piece.

Rubber Stamped Trapunto

Beading

Free Motion Thread Writing

Couching & Beading

scraps & metal

Part embellishing and part passport exploration, this page looks at 'scrap' from a different perspective; scrapbook store supplies, scrap bits of metal embellishments and the threads to stitch them.

Edge finish: fabric wrapped and attached with zigzag stitch

1 Cut a 3-3/4" square of nylon window screen. Stitch it on the middle of the base page with cotton, polyester, or upholstery thread. Carefully cut out a 2-1/2" square of copper mesh. Lay the mesh "on point" atop the nylon screen and stitch carefully. Use a straight stitch and cotton, polyester, or upholstery thread. Paintable wallpaper has a paper backing and is easy to sew through. Dab some paint on a 1-1/2" x 4" piece and stitch near the top of your base page with a light zigzag stitch.

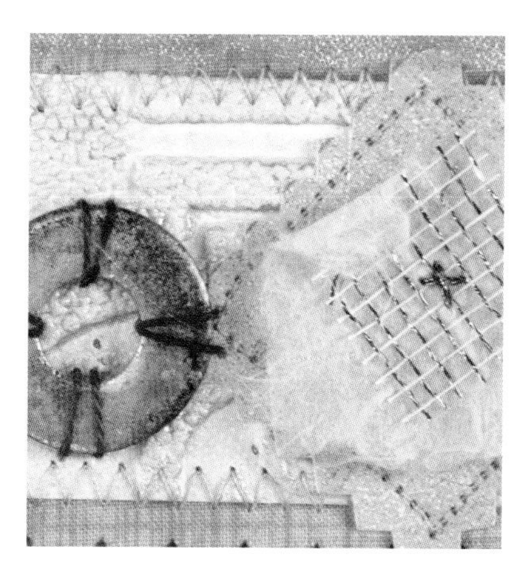

2 Couch metal washers in place with a few wrapping stitches using heavy weight cotton thread to securely hold the weight of the washer. Paint the washers with inks or nail polish before attaching if desired. Select three papers in different weights and fibers. Lightly iron a layer of fusible tear away stabilizer to the plain paper. Trim a 1-1/4" square with pinking shears. Stitch the base paper in place on the center of the wallpaper. Cut and layer the two remaining papers on top and tack in place with a cross-stitch in decorative thread.

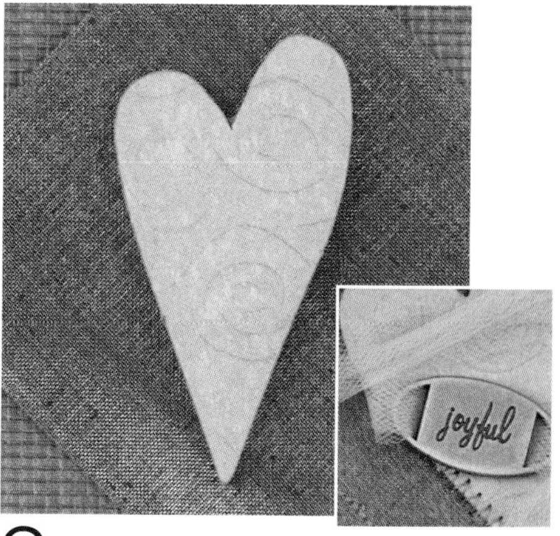

3 Ribbon slides are designed to slide over ribbon or strips of fabric. Grungeboard is a leather-like synthetic material similar in thickness to chipboard, but very stitchable. It comes in plain or embossed sheets, as well as a wide assortment of die cut shapes. Choose a Grungeboard die cut for your page and slide a 15" strip of tulle through a ribbon slide. Wrap it around the Grungeboard shape and lay in place atop the screens. Pin the tulle aside and stitch the Grungeboard in a few areas to secure it. Stitch a small line across the tulle at each side point of the copper screen to hold in place. This stitching will be covered by the brad. Use heavy thread that will not be cut by the screen.

4 To add brads, pierce the fabric page on either side of the copper mesh. Lay each loose end of the tulle over these spots and pierce with the brad to hold the tulle off to each side. When opening the brad at the back of the fabric, have the prongs open parallel to the sides of the page to keep them clear of the stitching. NOTE: Pin the tulle tails into the middle of the page to keep them clear of the edge finishing process.

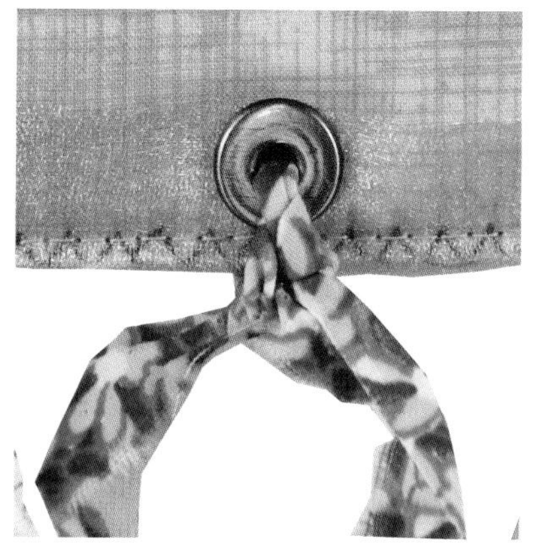

5 Referring to Edge Finishes on page 36, add an edge finish to the page and unpin the tulle. Mark spots for three eyelets approximately 1/2" from the bottom of the page and evenly spaced. Set the eyelets and tie 10" strips of your favorite fabric selvages to the page and let them hang below.

tips

- Use a slightly longer stitch length when working with paper. This will prevent perforations in the paper that may cause it to tear away.
- Thin metal and paper can be secured to your fabric with fusible web.
- Save used but still sharp needles for stitching through paper and metal. When finished throw these away.
- When stitching with metal, back the metal with fabric, felt, or a stabilizer to prevent scratches on your sewing machine bed.
- A sharp rather than ballpoint needle works best with paper and metal.

adding text

Whether you want to label a technique, add a sentiment, or express a theme, words can add depth and understanding to your pages.

Edge finish: lightly stitched with a zigzag and yarn couched around the edges.

getting started

There are many ways to add letters and images to your Passport page. Printable fabric sheets, stencils, and even decorative buttons will give your pages a voice.

Printed Tails are 18" - 20" strips of ribbon, lace, or fabric with words added to them. Print words directly onto a printable fabric sheet, following manufacturer's directions. Use colored pencils to add excitement to a boring black font. Work with the side of the pencil point to avoid sharp lines. Blend colors with a finger and press with a warm iron and paper towel to pick up any excess pigment. Cut the colored letters into individual pieces and stitch them to a pretty ribbon.

Transfer Artist Paper (TAP) is an easy to use transfer product for inkjet printers. Reverse the text before printing and follow the manufacturer's directions for transferring onto the fabric. Cut out the word, lay it over the fabric, and press for a few seconds. Remove the paper and stitch the transferred fabric to a tail made from a piece of velvet and rickrack.

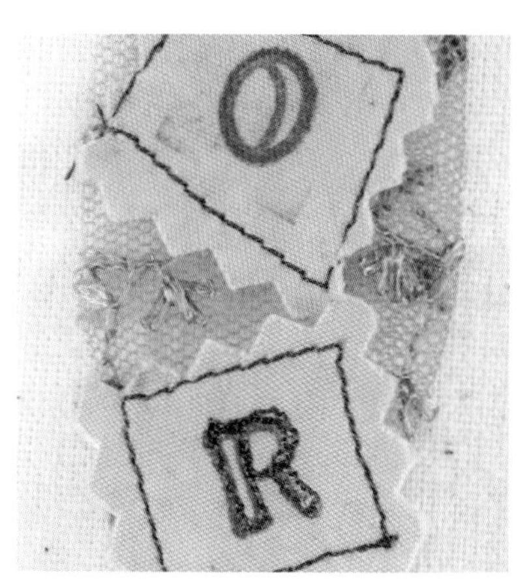

Refer to Rubber Stamping on page 170 before beginning. Stabilize the fabric with a fusible tear-away stabilizer. Ink your stamps one at a time and press each letter on the fabric. If desired, free motion stitch over each letter or leave as a simple stamped image. Trim the individual letters and stitch them to a tail of tattered lace.

Free Motion Stitching over a Foundation will give your free motion lettering a crisp look with a traced foundation. Lay wash- away stabilizer on top of the printed sample and trace the letters with a pencil. Consider tracing over twice so you have a "warm-up" word. Hoop a layer of tear-away stabilizer, the background fabric, and the traced stabilizer. Free motion stitch over the tracing. Remove from the hoop and cut away the excess top and bottom stabilizers. Rinse the fabric, allow it to dry and stitch it to the page over a small silk scrap.

adding text

Stencils can be used for reverse appliqué or for painting letters onto fabric. Back a square of fabric with fusible web. Leave the paper on for added stability. Trace the stenciled letter onto the fabric and carefully cut out the center sections, leaving the letter's outline. Hoop a piece of velvet or other fabric with stabilizer. Free motion stitch wavy lines in a variety of threads. Trim the fabric to a size that will fit just inside the stencil rectangle, amply filling the letter. Remove the paper backing from the fabric and fuse it on the stitched fabric. Use a narrow zigzag stitch to cover the raw fabric edges of the letter. Add a straight stitch over the zigzag if you need more definition. Stitch the joined piece in place on the Passport page with a decorative stitch. Add a band of decorative stitching to either side of the appliqué.

Printable Fabric, such as inkjet printable ExtravOrganza, adds a delicate touch. Trim the word and remove the carrier paper. Stitch in place on the page with your thread of choice.

Beads are a fun way to add text to your page. Stitch them on with beading thread or a heavy cotton or polyester in an arc below the appliqué.

Add an edge finish to your page referring to Edge Finishes on page
Place an eyelet in the upper corner the page. Join the three printed tails with a few cross-stitches and feed it through the eyelet. Tie the tails in a knot.

printing on ribbon

A unique way to include text in your work is to print it on ribbon using your inkjet printer. Add single words, favorite quotes, or make your own personalized quilt labels.

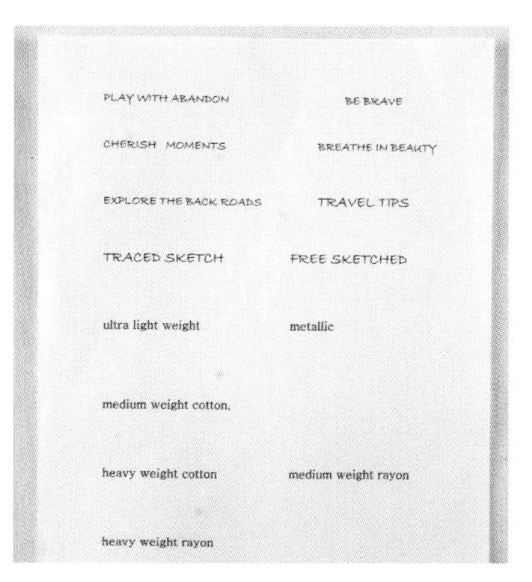

1 In your word processing program type out the text you would like to appear on your ribbon. Be sure to leave appropriate spacing between words you want to print individually so you can cut the ribbon between them. In addition, space the lines of text to allow for room for the ribbon. Print the text on a piece of standard printer paper. Do not close the word processing program.

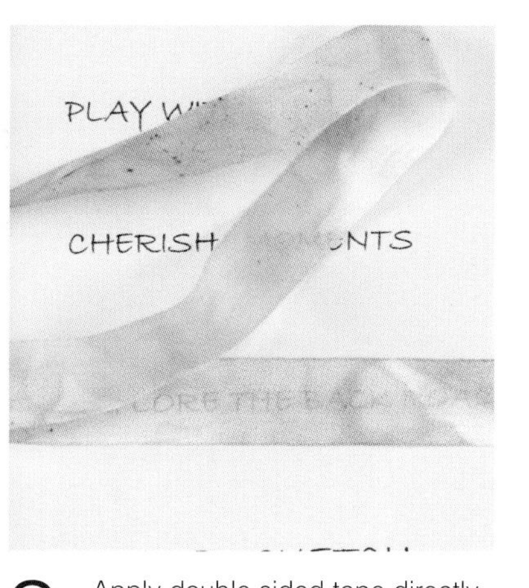

2 Apply double sided tape directly on top of the printed words. Smooth ribbon onto the tape. Do not stretch the ribbon as you place it. Place paper with ribbon back into your printer and print again.

Edge finish: picot beading

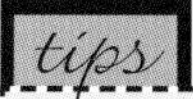

- To prevent fraying paint the ribbon's edges with a clear nail polish.
- Ink may bleed slightly on polyester and nylon ribbons.

adding pockets, stamps, & tags

Embellishment is all about decorating your space with a few pictures and objects d'art that speak of your style. Why should your Passport be any different? Stamp, stencil, and tag a page that is uniquely you.

Edge finish: zigzag stitch with ribbons woven through eyelets for a corseted edge

getting started

Stencils can be used to add images and graphics before, during, and after the Passport base page construction. Rubber stamps can be stand-alone embellishments or used as guides for thread painting, adding text, or as quilting motifs.

Stenciling: Lay your chosen stencil over the area you wish to paint. Dab paint onto a cosmetic sponge, foam brush, or stencil brush. Rub excess paint off on a paper towel. Lightly 'pounce' the sponge across the opening in the stencil. Carefully lift up a corner to check the image. When the image is complete, remove the stencil. Let the paint dry completely before adding any other embellishments to the stenciled area.

Rubber Stamping: Lay a stamp face up on the tabletop. Tap the stamp pad across the entire surface of the stamp. Place the stamp onto the fabric surface and apply even pressure for a few seconds with both hands. Hold the fabric down while you lift the stamp to cleanly separate them, as the fabric will sometimes cling to the stamp. Re-ink the stamp and add more images if desired. Allow the ink to dry completely before pressing or fusing. Use a pressing cloth or piece of paper to protect the image and blot any stray ink.

Adding Pockets: To make a pocket for your page, cut fabric twice the length of the finished height of the pocket and 1/2" wider than the page. This pocket is 2-1/2" tall, so the fabric is cut 5 x 5-1/2". Fold the fabric in half over a piece of 2-1/2" x 5-1/2" piece of batting or stabilizer to form a sandwich with the batting in the center. Stitch over the sandwich with a meander stitch. Embellish the pocket and add trim to the top edge if desired. Layer the Passport page base with stabilizer. Trim the pocket to the page width and pin in place around the bottom portion of the page. If desired, tuck a ribbon tab between the pocket and the page and pin in place. Stitch around the entire perimeter of the page with a zigzag stitch.

Add ribbon, paper, or fabric tags: Refer to Adding Text on page 166 and Printing on Ribbon on page 169 for instructions to add these embellishments to your page.

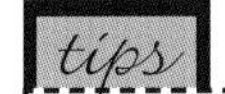

- Don't overload your brush with paint. It is always easier to add more paint then to try and remove it.
- Rinse the sponge and stencil immediately after use.
- Paint can be heat set with an iron if necessary.
- When choosing stamps, look for clean images and deeply etched stamps.
- Apply even pressure on the stamp.
- Work on a firm, flat surface.
- Stamp on single layers of fabric. Batting and other soft materials behind the fabric could spoil the crispness of the image.
- The fabric should be wrinkle-free.
- Pretest your fabric. Some fabrics have a coating or glaze making them resistant to taking a stamp image well.
- Stamp one image per inking. There will not be much ink left behind for a second image.

Lined Pocket Pages: Add fabric to the backs of pages before joining them. Leave the top of the page open to create a lined pocket page.

crazy quilting

Crazy quilts have traditionally been the canvas for decadent fabrics and over-the-top application of technique and embellishment. We lightened up our passport page with contemporary fabrics, machine stitching, and a sprinkling of handwork.

Edge finish: ribbon couched with zigzag stitch.

1 Ricing, page 96

2 Thread Painted Appliqué, page 86

3 Couching, page 90

4 Stitch Layering, Programmed Stitches, page 23

5 Ribbonry, page 150

6 French Knots by machine with Heavy Threads, page 58

7 Beading, page 160

8 Ribbonry, page 150

9 Ultra Heavy Thread and Bobbin Stitching, page 58; French Knots, page 95

10 Rubber Stamping, page 170, highlighted with Hand Stitching, page 94

11 Make Your Own Fabric, page 128

12 Thread Lace Grid, page 110; Thread as Structure, page 122

1 To create a base for your stitches, use at least three different fabrics to cut several strips approximately 10" long and varying the width between 1" to 2". Piece the strips together and press the seams to one side. Draw a curve shape approximately 4" wide on the pieced strips and cut out. The curve should fit within the width of the page with space on each side. It will be longer than the page and will be trimmed later.

2 Stitch a ribbon down the length of the background page if desired. Place the curved fabric piece on the background page and stitch with an open zigzag stitch down either side.

Add a variety of stitches, as well as other techniques you've already learned, to the Passport page. Refer to page 172 for a chart showing the techniques and stitches we used.

Give your passport page a more traditional Crazy Quilt look by surrounding it with foundation-pieced strips of funky fabrics, silks, or velvets. Embellish it with stitches, beads, and yo-yos.

Making Yo-Yos & Buttons

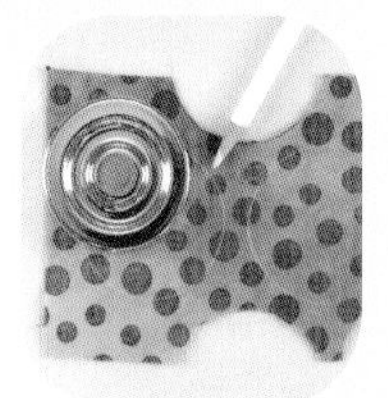

1. Trace a circle onto fabric and cut out.
2. Knot a sturdy cotton or polyester thread and create a running stitch 1/8" inside the perimeter of the circle.

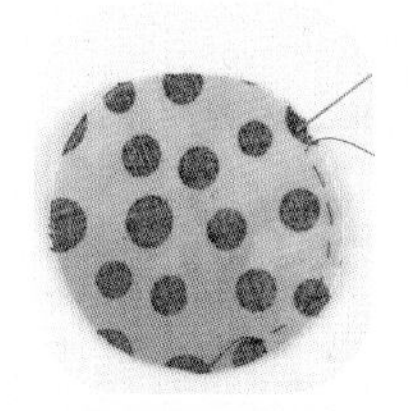

3. At the last stitch, gather the fabric circle into a 'poof' and tie off the thread.
4. Attach the yo-yo with tacking stitches and embellish the center with bead dangles.

To make your own custom covered button, thread paint an image within the circle of fabric. Follow the manufacturer's instructions to finish the fabric covered button.

resources

We highly recommend you look first to your local quilt or fabric store to build your thread stash. Thread manufacturer web sites often have a great deal of wonderful information, tips, projects, and inspiration. We have listed the threads we used in the book, as well as a few of our other often used products.

American & Efird
(Robison-Anton, Signature, Mettler)
P.O. Box 507
Mt Holly, NC 28120
800.438.5868

Aurifil Threads
500 N. Michigan Avenue
Chicago, IL 60611
www.aurifil.com

The Caron Collection
55 Old South Ave
Stratford, CT 06615
203.381.9999
www.caron-net.com

Coats and Clark
(including Star Quilting Threads)
P.O. Box 12229
Greenville, SC 29612-0229
800.648.1479
www.coatsandclark.com

Gutermann Thread
PO Box 7387
Charlotte, NC 28241
704.525.7068
www.gutermann.com

Kreinik Threads:
P.O. Box 1966
Parkersburg, WV 26102
800.537.2166
www.kreinik.com

Madeira Threads
Tacony Corporation
1117 Gilsinn Lane
Fenton, MO 63026
www.sewingandcraftclub.com

Marathon Thread
1050 Colwell Lane, Unit 204
Conshohocken, PA 19428
800.984.7323
www.marathonthread.com

Linda Palaisy hand dyed threads
www.lindapalaisy.ca

Oliver Twist Threads
Available through www.joggles.com

Presencia
PO Box 2409
Evergreen, CO 80437 US
866.277.6364
www.presenciausa.com

Stef Francis (U.K.)
www.stef-francis.co.uk

Sue Spargo-Folk art quilts
1364 Walnut Ridge Dr
Uniontown, OH 44685
www.suespargo.com

Sulky
980 Cobb Place Blvd.
Kennesaw, GA 30144
800.874.4115
www.sulky.com

Superior Threads
87 East 2580 South
St. George, UT 84790
www.superiorthreads.com

Thread Gatherer
2108 Norcrest Drive
Boise, ID 83075
208.387.2641
www.threadgatherer.com

Valdani Inc.
3551-199 ST.
Edmonton, AB
T6M 2N5 Canada
www.valdani.com

Weeks Dye Works, Inc.
1510-103 Mechanical Blvd.
Garner, North Carolina 27529
www.weeksdyeworks.com

WonderFil
Bay 3, 2915 19th Street N.E.
Calgary, AB
T2E 7A2 Canada
www.wonderfil.net

YLI Corporation
1439 Dave Lyle Blvd.
Rock Hill, SC 29730
www.ylicorp.com

ON-LINE THREAD RESOURCES

E-Quilter
www.equilter.com

Joggles
www.joggles.com

Red Rock Threads
www.redrockthreads.com

The Thread Studio (Australia)
www.thethreadstudio.com

NON-THREAD RESOURCES

Artistic Artifacts
www.artisticartifacts.com

Jacquard/Rupert, Gibbon & Spider, Inc.
www.jacquardproducts.com

Transfer Artist Paper by Lesley Riley
www.ctpub.com

The Crop-A-Dile Eyelet & Snap Punch Tool
www.weronthenet.com

WireMesh & Art Emboss Metal sheeting
www.amaco.com

Tim Holtz Grungeboard
www.timholtz.com

Golden Threads Quilting Paper
www.goldenthreads.com

The Crafter's Workshop Stencils
www.thecraftersworkshop.com/

Kunin Felt
www.kuninfelt.com/

Fiber Goddess
www.fibergoddess.net/

Renaissance Ribbons
www.renaissanceribbons.com

Schmetz needles
www.schmetzneedles.com

Tattered Angels Glimmer Mist
www.mytatteredangels.com

Tsukineko- VersaCraft Inkpads,
Markers & All Purpose Inks
www.tsukineko.com

STABILIZERS:

Ask for these stabilizers
at your favorite quilt shop.

- Aqua Film
- Dissolve
- Flexi Firm
- Heavy Weight Tear Away
- Lutrador
- Madeira
- Magic Film
- Peltex
- Solvy
- Timtex
- Totally Stable
- Wet-N-Melt

FUSIBLE WEB:

Ask for these fusible webs
at your favorite quilt shop.

- Fast 2 Fuse
- Fine Fuse
- Heat N Bond
- Misty Fuse/Attached Inc. www.mistyfuse.com
- Steam A Seam
- Stitch Witchery
- Wonder Under

Acknowledgments

A book may be written by few but many are involved in its production. We would like to thank Aurifil, Coats and Clark, Gutermann, Kreinik, Linda Palaisy, Madeira, Marathon, Presencia, Sue Spargo, Superior, The Thread Gatherer, Valdani, Weeks Dye Works, WonderFil, and YLI for contributing thread for our research and creative exploration. We are grateful to Art Gallery Fabrics, Baum Textiles, Northcott Fabrics, and Westminster Fabrics, for their generous donations of fabrics to help our stitches shine.

Thank you to all the thread artists who walked this road before us and generously shared their experience, talents and insights to ease our journey.

Last but not least we would like to thank our wonderful team at Landauer Publishing for giving us this opportunity to share our passion with thread enthusiasts everywhere.

about the authors

To fairytale beginnings, friends I have found in every class and especially my family; you hold my heart and I am forever grateful.

Debbie Bates

Debbie Bates is a fiber artist who loves to stitch. Early classes in traditional quilting and a penchant for wanting to create original work unlocked doors to her stitching addictions. Sharing anything from the ordinary to subversive has become her calling.

Debbie has been teaching in the world of fiber arts for over 1 years. She lectures and teaches across North America, writes about stitching and the creative process, and works in the sewing industry. She has shown and sold her work in Canada and the United States. She is a regular contributor to many publications and has appeared on a variety of television programs. History, mythology, travel, and the surreal all play a part in Debbie's work. She is strongly influenced by music, movies, and other art that speaks to life's undertones.

Debbie lives in Ontario. Her four children are on their own paths of discovery from home and afar. Her faithful pup, Budc waits in the window for her every day and her husband Phil loves and supports her through it all. (The cat does not seem to care.)

Debbie shares her creative discoveries and stumblings in her blog, http://stitchtress.wordpress.com or website, http://www.stitchtress.com

To the Breck Girls—your love, support and friendship mean the world to me.

Liz Kettle

Liz Kettle is a fabric and mixed media artist with a passion for teaching others the joy of making art and the creative process. After filling her tool box with the skills for success in the traditional quilting arena she began to delve into art quilting and discovered a world of freedom and fun in mixed media.

Liz's eclectic work is influenced by her beautiful surroundings, the foothills of the Rocky Mountains, and her love of vintage textiles and history. She incorporates layers and found objects to tell stories of the land and people. When she isn't creating visual art, Liz writes about creativity and the creative process. She is the founder of Textile Evolution, a unique retreat (www.textileevolution.com).

Liz lives in Colorado with her very supportive husband. Her three amazing sons are beginning their own life journeys and her two grandsons are keeping everyone on their toes.